Questions and Answers About
PAY TV

by
Ira Kamen

TEXAS STATE TECHNICAL INSTITUTE
ROLLING PLAINS CAMPUS – LIBRARY
SWEETWATER, TEXAS 79556

HOWARD W. SAMS & CO., INC.
THE BOBBS-MERRILL CO., INC.
INDIANAPOLIS · KANSAS CITY · NEW YORK

FIRST EDITION

FIRST PRINTING—1973

Copyright © 1973 by Howard W. Sams & Co., Inc., Indianapolis, Indiana 46268. Printed in the United States of America.

All rights reserved. Reproduction or use, without express per-
mission, of editorial or pictorial content, in any manner, is
prohibited. No patent liability is assumed with respect to the use
of the information contained herein.

International Standard Book Number: 0–672–20971–3
Library of Congress Catalog Card Number: 73–79073

Foreword

Broadcasting and electronic communications have been my absorbing professions for over 50 years, including the gradual growth of pay tv.

The author of *Questions and Answers About Pay TV* has been a friend and associate for over thirty years, and I have served in many trenches with him in the war against those who oppose change.

The book tells how it is, with the sweet perfume of reading environment that is created by a knowledgeable writer who is in the pay tv business and answering questions on a daily basis.

Ira Kamen has colored some factual answers with the aroma of his dedicated belief in the American system and viewer power. For over twenty years, the interests, who have sat on the sidelines, making no effort to nurture the development of pay tv, have now come to realize that this form of selected entertainment will bring low-cost quality programing to the mass audience.

The story which evolves in this question and answer format is that financiers, producers, and technical talents who have created ingenuous products, and the FCC, are now allied to give the viewer the power in television to buy what he wants, from whom he wants, in the comfort of his home. The freedom of choice between pay tv and advertiser-sponsored television is for the first time, really free television. *Questions and Answers About Pay TV* describes a new milestone of growth for American communications in the public interest.

JOHN R. POPPELE

John R. Poppele is a fellow of the IEEE, and formerly director of the Voice of America. He has been President of the Broadcast Pioneers, FM Broadcasters Association, and President of the TV Broadcasters Association.

Mr. Poppele is currently President of Tele-Measurements, Inc., President of the Veteran Wireless Operators Association, and Vice President of the Radio Club of America.

801756

Dedication

To my late beloved partner, Daniel J. Riesner, with whom I rubbed minds for ten years to spark the adventures which are now Theatre-VisioN pay tv systems, Educasting Systems, Comfax Communications, ComputerPix, Laser Link, and a host of acquisitions and mergers which remain as a tribute to his memory.

IRA KAMEN

Acknowledgements

The author sincerely appreciates the cooperation that pay tv proponents, Federal Communications Commission representatives, and program sources gave to his market researcher, Kathy Doane.

Questions were posed by the pay tv proponents to best bring out the facts on their systems, plans, and policies. The list of those who helped is too long to put into a credit note. This book could not have come into being without the creative thinking of the businessmen and engineers who were interviewed.

Specific credit for reviewing the Rules and Regulations is given to FCC Counsel William Bernton, of Mallyck & Bernton, and to pay tv pioneers Will Baltin and Jack Berens, who reviewed sections of this text to assure its accuracy.

Contents

Chapter 1

Introduction to Pay TV

What is pay television?

Pay television is a new service which will soon be offered to the American public. It will bring, to a special subscriber audience, both over-the-air and cable broadcasting of the latest motion pictures, sporting events, and special entertainment which are not broadcast by regular services because of high cost or contractual problems. Pay tv makes it possible to broadcast these otherwise unavailable attractions free of annoying commercials by allowing the viewer to pay for each program that he watches in his home, just as if he were buying a ticket. Obviously, when a viewer pays for what he watches the high cost of talent and production can be recouped in the same manner that the theatre owners make profit for producers. However, as will be explained later in this text, pay tv will restore an audience to the motion picture industry which has been lost to film producers.

The deal is a good one for the viewer. True, he has to pay for entertainment instead of watching free, but he only pays for those attractions which he sees. The rest of the time he can watch the usual advertiser-sponsored shows. But those shows he does pay for, he enjoys in the comfort of his living room, without moving a step, with his family and friends, and where the refrigerator is handy. When the subscriber is at home, seeing shows and sports that he would otherwise have to make a special trip to attend, he is saving the cost of travelling to and from the entertainment center. He also avoids baby-sitting costs and he trades an uncomfortable seat in a theatre for a comfortable seat in his home.

You cannot put a show on the air or cable and trust people either to pay or to turn to another station; therefore, special transmission and reception methods are used for pay tv. The picture and sound are sent out over the air or on cable in such a way that ordinary tv sets, with a special device, allow the viewer to watch and enjoy the program. On sets without this device, the picture may be unstable, the colors may be reversed like a photographic negative, the scenes may flip-flop, the picture may be jammed, or the sound distorted. Many other methods, to code the program and ensure that those who are not subscribers cannot watch the program for free, will be described in this text.

Each subscriber to a pay tv system gets a small decoder/tuning device which is installed between the antenna and the set. He is not required to have any kind of special set. The device must, by FCC Rules and Regulations, install between the antenna and the set without disturbing the regular performance of the set for free television. After installation, he can watch one of the special programs by simply flipping the switch to the pay tv program and by using the stimulating device, which may be a ticket, a card, a telephone connection, or other devices which will be described in this report. The decoder then clears up the picture for comfortable enjoyment of the program in the home. Pay tv is not being proposed as a substitute for advertiser-sponsored tv. It is being offered as a lower-cost method of providing better entertainment for the American public in the safety of their home. All of the secondary costs (driving, parking, babysitting) are eliminated, and usually for the price of no more than a single ticket.

When was pay tv first proposed?

Pay tv was proposed in 1950 by the Zenith Corporation, whose first system was called Phonevision. Phonevision was tested in Chicago in 1951. This initial system was over-the-air pay tv and it has had a stormy past.

The history will not be recounted here, except to note that it was initially opposed by the motion picture theatre interests who were, for a while, able to elicit significant support from Congress. In the early 1960s Zenith Radio Corporation, along with RKO General, fought another battle with the Connecticut Motion Picture Exhibitors in which they defeated the Exhibitors. The U.S. Court of Appeals, for the District of Columbia Circuit, held that the Federal Communications Commission did have the power to authorize a pay tv operation in Hartford; therefore, the action in granting that authorization was within the scope of the FCC's authority. The theatre owners sought to appeal this decision to the United States Supreme Court, and the Court refused to entertain the appeal.

Who has tested pay tv and what have they proven?

Tests have been made in various areas of the North American continent by Paramount Pictures in Etobicoke, Canada (a suburb of Toronto), in conjunction with Famous Players; RKO General and Zenith in Hartford, Connecticut; Columbia Pictures and Jerrold in Bartlesville, Oklahoma. Attempts were made to install systems in Los Angeles and San Francisco by a corporation, under the leadership of Matty Fox and Pat Weaver, called Subscription TV, Incorporated (STV). Attempts were also made by Teleglobe Pay TV in Denver, Colorado.

The tests made in Etobicoke via cable, and the over-the-air tests made in Hartford, proved that people would pay on a per-program basis for pay tv programing. The Bartlesville test was a cable system that charged a monthly rate for all the motion pictures available via the subscription tv company. It failed because the company could not obtain better program material, and people would not take an assessment for pay tv programs that they did not see. Additionally, producers would not release program material that was not paid for on a per-program basis. The test in Los Angeles met with what was perhaps the most fantastic action of a vested interest to prevent a pay tv service. The theatre operators managed to pass a referendum in California against pay tv. By hawking the antipay tv story in newspapers and at their box offices, they persuaded the citizens of California to vote against the pay tv service in a referendum at the polls. This halted a $25 million effort, and the entire operation collapsed. Court rulings subsequently indicated that this was unconstitutional. If the same logic were used, the glue manufacturers could have put up a referendum and prevented the sale of plastic adhesive tape. The theatre owners, by a magnificent effort, destroyed this test. It did have the potential of success; deals had been made to present the many forms of entertainment possible with pay tv. The Teleglobe test that was approved for Denver was a practical system, but the motion picture operators were not ready to supply the films. The financiers behind the Denver test (Bartell Media) did not have sufficient capital to take on the negative cash flow which would have been needed to launch this test, subsequent to the collapse of STV.

None of these tests, however, could have been converted into profitable operations because the time for the market had not really arrived. The hurdles created by the opponents of pay tv, the under-development of cable tv systems in the early 1960s, and the general conditions of the marketplace were not in support of the development of pay tv.

As we will explain in the series of questions and answers in this text, the marketplace has developed, the authorizations from the FCC

9

are now clearly defined, and there are presently enough major proponents to make pay tv work. The motion picture industry now agrees that if they are to become the great producers of real entertainment for the home, they must join with this new medium.

Why is pay tv so closely related to the development of urban cable tv?

Cable tv is proposed as a function which could provide many new services for the home. That is, services which will enable the consumer to buy products, interrogate information banks, and many other sophisticated educational and business services. This all may be true, but a subscriber in an urban area, who is receiving good reception from the three networks and a couple of independent stations, must be highly motivated to join the cable system. Urban cable tv, which uses buried cable, is the most expensive system for the cable operator to install. If the cable operator is to successfully develop urban cable television, with its classic and avant-garde services and access channels, it must first develop subscribers. To obtain these subscribers the operator must provide desired entertainment. Subscribers must get first-run films or other entertainment that they would ordinarily pay for by going to an outside area. It will take years before people learn the value of some of the programs which are planned. For example, to consider the home a polling place, one must first get the two-way device into the home.

If people are to use a decoder/tuner unit, with response buttons, it must first be installed under economically sound conditions. It has been estimated that the average terminal which will go into the home will probably cost in the neighborhood of $100. In a five-year straight line write-off, this means that $20 per year has to be written off to provide a device which will meet all of the Federal Communications Commission's specifications. It is important to remember that in selling cable service where improved reception is not the criteria, first-run films and other desirable entertainment offered by pay tv should be emphasized.

Why is pay tv more likely to succeed at this time than at any other?

All conditions are favorable for pay tv. First, the producers want it to regain the market which they have lost as a result of changing conditions. These conditions being that in 1972, 90 percent of the market for film presentations were people under the age of 30. The consumer wants the service. There is a projection that by 1980, there will be 77 million television households. The increase in the crime rate in the center city makes those who live there afraid to go out at night. This sensitivity to going out at night is also becoming very high in the suburban areas. Records have shown that a large body of middle-class Americans prefer to stay home rather than go out. They have been staying away from theatres in droves, especially those who are mature or who have the need for babysitters.

How big a market can pay tv be?

As stated previously, there will be over 77 million television households in the United States by 1980. If each of these families were to spend approximately $10 a month on home entertainment, they would develop a market that would approach $10 billion. If three quarters of that revenue were for watching the better motion pictures at home, the film producers (who might collect 45 percent of this amount) would increase their annual volume to $4½ billion. This would be several times their return from the 1972 box office receipts, which was only slightly over $1 billion.

What will be pay tv's biggest year?

In 1973, the tests conducted by most of the proponents will have been completed, and all of them will have tooled for production in 1974. The steep rise in the curve of pay tv growth will begin in 1974. It should stay on a steep curve of accelerated and explosive growth until at least 1980.

Will the growth be on over-the-air or cable television?

The growth of pay tv will be in both areas, with cable television and the "wired nation" offering the greater potential. However, the over-the-air services will provide instant subscribers since it will be unnecessary to wait for the development of cable tv. At some future date it will be possible to interface and convert the over-the-air signals so that they can be provided for use with the over the cable television system.

How will the conversion of over-the-air signals to cable be accomplished?

It will be done by assigning a specific channel to the over-the-air pay tv company and having one master decoder at the head end. This decoder will convert the over-the-air type of decoding to cable tv encoding. If done with careful engineering, all the rules which apply to over-the-air pay tv can be applied to the cable television performance, and there should be no degradation in picture due to this double conversion.

Where is over-the-air pay tv likely to start?

Zenith Radio Corporation has filed requests for both Chicago, Illinois, and Los Angeles, California. As Zenith is an equipment manufacturing company, and is believed to be tooled for the pay tv equipment it developed after a six-year experiment, it is assumed that they will enable the industry to start in 1974. Teleglobe should be operational next in San Francisco, California, with Blonder-Tongue in Newark, New Jersey and TheatreVisioN following shortly thereafter.

Are any of the cable operators functioning with the total service in a major market?

Yes, TelePrompTer has been offering a pay tv service in their Long Island installation at Babylon Village and in parts of the town of Islip. Via Suffolk Cablevision, TelePrompTer operates systems in Islip, Patchogue, Brookhaven, Smithtown, and Bayshore. They are offering a converter with a special channel "P" which will give any of the 25 thousand or more subscribers a package of basketball and hockey games at a cost of approximately 65 cents per game. This system is not a per-program system, but one must subscribe to the total package in order to get the special channel selector with this new channel "P." TelePrompTer anticipates that this seasonal package will be priced at approximately $50. The total package includes the Nets basketball team and the Islanders (a local hockey team), neither of which are broadcast on free television. The Long Island Coliseum promoters ruled out, at the beginning, any free televising as it would have an adverse effect on box office receipts. The total of eighty games played by the Nets and Islanders works out to approximately 65 cents per game for home viewers. As an incentive, the TelePrompTer organization would make post-season play-offs free to the subscriber. In this arrangement, it was not considered necessary to have an encoding and decoding system. While the purchase of a special converter would make some of these channels free to knowledgeable subscribers, it was doubtful that there would be meaningful cheating. No such cheating had been reported at the time of this publication.

Why are some systems only designed for the hotel/motel market?

There are two reasons: First, there are organizations who lease television sets, and who are organized to serve only this hotel/motel market. Some pay tv system groups believe that there is greater revenue from the hotel/motel occupant than from residential markets. Rooms rented in the center city have occupants who are afraid to go out at night. The entertainment, therefore, is either regular television or the better programs of pay tv, which are uninterrupted by commercials. To benefit from this market, a number of proponents have designed their equipment only for the hotel/motel market. They have added, in many cases, features which are attractive to hotel operators. In early 1973, there were hundreds of hotels which were committed to having pay tv installations made for the benefit of their guests. By 1974, it is anticipated that thousands of hotels and motels will be committed to providing this new service. All the surveys taken during the tests of the various proponents indicated that pay tv was, indeed, considered a blessing by those who were confined to motel room living. This is true for the individual salesman as well as the family.

Do television viewers want pay tv?

A nationwide poll revealed that 43 percent of all persons interviewed would like to give pay tv a try. The poll indicated that only one out of every three viewers considered conventional tv good or excellent; while one in four considered conventional television poor. Approximately nine out of ten people believe that pay tv should be an authorized service.

What was the reference information which the FCC used to base its decision to go ahead with pay tv?

Most of the information on which the FCC based its decision to go ahead with pay tv came from the Hartford, Conn., test. This was a six-year operation (from 1962 to 1968) conducted by Zenith Radio Corporation's pay tv Phonevision system in that city.

What did these tests prove?

The tests showed that 41 percent of the families who enjoyed pay tv earned between $4 thousand and $7 thousand a year from 1962 to 1968. Pay tv was a low-cost form of entertainment since they did not have to pay for babysitters, parking, and the high price of motion picture tickets. With pay tv, they paid less than the price of a single ticket for the whole family to watch a top-notch box office attraction. It also showed that 43 percent of the people who used the service earned between $7 thousand and $10 thousand per year; therefore, pay tv was not a service desired only by those in the upper income brackets. In fact, hospitals, nursing homes, and rest homes offered very favorable reactions to enjoying box office attractions at a minimum cost for entire groups. Access to theatres in bad weather, the high cost of a single admission, etc., would have prevented these groups from enjoying better films.

During the Hartford tests, how much did the average subscriber pay to view one program, and how many programs did he view?

The average subscriber, according to the report filed by Zenith, spent approximately $2 a week to view one program.

Is there a national consumer group that reported favorably on pay tv?

Yes. The National Consumers League (founded in 1899) serves the nation's consumers in Washington, D.C. They have stated that they favor "affording the television consumer the freedom of choice to enjoy subscription television and, therefore, [they] oppose efforts directed at restricting this choice."

Who are the designers of pay tv systems?

While most of the designers of pay tv systems are staff members of the pay tv proponents, there are some consultants and designers who are also product suppliers to the industry. At this writing, the designers for several of the systems are: Donald Kirk and Michael Paolini from Digital Communications of St. Petersburg, Florida. Kirk and Paolini did some of the basic design work for the Home Theatre Network system for Gridtronics and for Computer Television. The Laser Link Corporation is designing products for the TheatreVisioN Corporation and for the Philco-Ford motel systems.

Who are the suppliers of some of the production products?

Thompson Ramo Wooldridge is a supplier of Optical Systems equipment. Philco offshore manufacturing operations may be a supplier and Data-Systems, of Columbus, Ohio, are the suppliers of Home Theater Network equipment. The Oak converter people of Crystal Lake, Illinois, (K'Son makes Columbia's Tele/Theatre module) are suppliers to the Columbia Pictures Tele/Theatre and other operations. As the market advances, there will obviously be more offshore capabilities used, and relationships will be organized between pay tv proponents and suppliers. In the over-the-air field, the manufacturers are Zenith Radio Corporation for the Pay Television Corporation program and Blonder-Tongue, acting as its own equipment manufacturer. Teleglobe and TheatreVisioN will probably subcontract.

What is the patent position of the pay tv proponents?

Nearly all of the pay tv proponents have proprietary patent positions. This is especially true of over-the-air proponents. One or two of the cable system developers appear to be willing to sublicense in order to accelerate the growth of the cable television industry.

What major cable pay tv systems are in operation as of early 1973?

The first full-blown system, in which 1000 units are installed in the homes and 500 in hotels, is the TheatreVisioN system, which is located in Sarasota, Florida, using the Storer Broadcasting Cable Television facility. This system has been designed for expansion to the cable systems and franchise areas in the southwest of Florida. Home Theater Network will be offering its system to the cable television subscribers in Redondo Beach, California. Optical Systems is now in San Diego and Gridtronics is in ten communities. Athena is in Jefferson City, Missouri. Home Box Office's system is located at Wilkes-Barre, Pennsylvania; Cinca Communications is in Long Beach, California. Others have experimental systems as described in this text.

How is pay tv expected to develop from the areas described above?

It is assumed that it will develop rapidly, with those who have hardware available in control of the situation. All cable tv systems operating, or being planned, are studying the availability of pay tv to be offered to its subscribers. There is not a pay tv proponent who has not contacted all of the major multiple system operators (MSO's) in the country.

Which is more important to pay tv, hardware or software?

Hardware will be in control in the beginning. When the industry stabilizes and production products are available from everyone, it is anticipated that software will control the market.

Are any of the pay tv proponents planning to develop software?

Yes, TheatreVisioN's president, Mr. Dore Schary, the former head of Metro-Goldwyn-Mayer and RKO pictures, has currently bought foreign software. The corporation is planning to obtain properties which it could convert into motion pictures under its own auspices, or as joint ventures with others. Time-Life is in a similar category, and it is assumed that others will realize that at some plateau they will have to contribute to the software package. Home Theater Network is video taping proprietary program material which it may offer to pay tv proponents with in-place systems.

How does pay tv help the community where it is installed?

The cable operator must have some highly profitable service if he is not to be faced with losses from the high risks he takes in providing channels for nonprofit making services. Many cities are being assisted by the Mitre Corporation, by Urban Information Service, and by other specialists in advanced technology. Those who are awarded franchises are burdened with the promise of providing all the capability of a wired city, in addition to just cable television. The cable operators will need the high income that they may be able to generate from pay tv in order to provide the services which will take a long time to develop. All those who have evaluated the services realize that they are going to be hampered by the limited knowledge to some people and by the lack of understanding of the possibilities of special services other than pay tv.

How does pay tv help the CATV franchise holder who has not yet been in operation?

There are many areas where people received their franchises in the mid or late 60s but did not move to build a system. In most cases, the cable company argued it did not have to proceed because of the FCC freeze which existed until March, 1972. This FCC freeze did not permit the cable company to acquire the distant signals needed in order to make the service more desirable for potential subscribers. Some of the cable companies, who had franchises in areas already served with Grade A signals from the three networks, were seriously concerned about bringing in a couple of independent channels from a great distance. They were afraid this would inhibit the accelerated development of subscribers and, therefore, block the cash flow needed to bring forth a profit-making business. Actually, the advent of pay tv with original motion pictures is a greater subscriber incentive than importing programing of nearby independent vhf or uhf channels.

What is the attitude of the financial community toward pay tv?

The Wall Street Community Investment Bankers, the venture people at insurance companies, and the bank specialists in cable tv believe that the development of urban cable television will be dependent upon the success of pay tv.

Why have they come to this conclusion?

Market surveys have shown that the special services proposed for leased channels and other services (including two-way networks) will not make a cable system viable in the urban community, even if they import the independent channels from other areas. Many of the reports of the financial community have indicated that there must be the incentives of pay tv programing if there is to be any motivation for the home viewer to subscribe to the service.

Does the cable industry feel the same way as the investment community?

Yes. This is the reason why so many multiple systems operators are proponents of their own pay tv system. Some have privately stated that they are flirting with financial disaster if they cannot provide pay services in major markets. They accepted the financial responsibility of providing technological innovations from which they have no assured income. Pay tv fare is the only high-income service that they can anticipate. All other services, including the experiments with unproven two-way gear, the high cost of underground construction, and digital communications represent high risk factors; they may not generate any income at all.

Is there a possibility that CATV will become a utility and that pay tv will effectively have to sell its services to a public utility?

Cable tv is a medium that informs, entertains, and influences its subscribers. This is not true of essential services like power, telephone, and water. While some cities have considered operating cable tv themselves, it is not considered politically wise to risk tax dollars and public criticism in order to finance an experiment that is as complex as cable television.

At the point of this writing, making cable television a public utility has been attacked from all sides, and it is assumed that pay tv will not be a service which has to negotiate with public utilities.

As Canada is even more saturated with cable television than the United States, what is their attitude toward pay tv?

Nearly all of the big cable television companies in Canada are planning to incorporate pay tv into their systems.

Are the Canadians only developing pay tv or are they, too, writing specifications which cover the wired city?

Just as in the United States, Canada is planning to develop new communications technology and to increase community dialogue. However, in some instances, the Canadian Department of Communications is planning to fund some of the developments of special services. Canada has been working on several projects to meet the need to deliver promised community involvement and has had good results with certain isolated groups. They have found that, through the process of creating new communications ties, new jobs have been opened, new coalitions have been formed, and new avenues of communication have been stimulated.

In 1973, numerous announcements have been made that Canadian hotels will have pay tv, and more than a dozen Canadian cable operators have indicated that they plan to offer pay tv to their cable subscribers.

What problems could affect the growth of pay tv?

It is known that President Nixon is opposed to pornography and permissiveness. Should the Federal Communications Commission determine that R-rated films cannot be transmitted by microwave or over cable, it could stunt the growth of this new medium. To quote Clay Whitehead, the head of the President's Office of Telecommunications Policy:

> Anyone who undertakes to use the electronic media to intrude into the privacy of the home with anything that is excessively violent, obscene, or with things that are directed at children and are damaging to the development of the moral character of children certainly should not be doing what they're doing. I think it's perfectly appropriate for the Congress and the FCC to adopt measures to be sure that the privacy of the American home is not invaded.

In early 1973, Senator John Pastore, head of the Senate Communications Subcommittee, also made similar statements. This might lead to a battle which could end up in the courts. Any censorship imposed would put pay tv proponents into a dilemma, since pay tv must rely, in the main, on feature films as entertainment for pay purposes.

How about opposition from theatre owners?

The Federal Communications Commission makes rules in the public interest. Their last ruling against the theatre owners clearly indicated that they will not support vested interests versus their charter to provide the American people with the maximum services in the frequency assignments.

Chapter 2

Rules and Regulations for Pay TV

What are the Federal Communications Commission guidelines governing over-the-air pay tv?

The FCC's Fourth Report and Order in Docket 11279 is the bible for over-the-air subscription television service. It is published in full in Part II of the Federal Register of December 21, 1968. That document did, however, leave unresolved a number of questions relating to technical standards and to procedures for applying for authorizations to engage in pay tv. These matters were disposed of in the Fifth Report and Order of September, 1969, in the same docket in which Sec. 73.644 of the rules (technical standards) was promulgated.

In what situations will pay tv be authorized?

The rules promulgated in the Fourth Report and Order provide, in general, as follows:

General Regulatory Provisions (Sec. 73.642)

1. Pay tv authorizations are granted only to television stations, not to third parties who contract with stations. Like station licenses, such authorization must be renewed every three years.
2. Pay tv will be authorized only in communities receiving Grade A service from at least five commercial television stations, including the station applicant. Only one station in each community will be given a pay tv authorization.

What are the station's responsibilities in a pay tv system operation?

1. A television station applying for a pay tv authorization, or the renewal thereof, must report to the FCC the full details of all its arrangements with program suppliers.
2. The station is responsible for the suitability of all programs carried. It must also retain the right to preempt any time segment if, in its judgment, the time is required to carry a nonpay program of local or national importance.
3. The rules state that the station should retain the right to schedule all pay tv programing and should not enter into any contracts which restrict its choice of program suppliers. However, the rules also provide that both of these requirements may be waived by the FCC.
4. The station must retain the right to set the maximum fee to be charged for any program.
5. Pay tv must be made available to all persons within the station's Grade A contour who receive a satisfactory signal from the station and who desire it. Charges, terms, and conditions of service must be uniform to all subscribers. Because of the likelihood of obsolescence, decoders must be leased to the subscribers; they may not be required to purchase them.

>   NOTE: The FCC recognized that various interests would be involved in pay tv, such as equipment manufacturers, program suppliers, and stations. It also recognized that, in some instances, several (or even all) of these interests might be centered in one party. As noted previously, it reserved great latitude in dealing with the interrelationships of these interests.

What are the pay tv programing requirements?

1. There are complicated "antisyphoning" rules designed to prevent pay tv from "syphoning off" feature films and sports events that are an important part of free television programing. The details, which can be found in Sec. 73.643(b)(2) of the FCC rules, will not be set forth here, except to note one of the principal provisions. This requirement prohibits the exhibition of feature motion pictures on pay tv at any time after two years following the general release date anywhere in the United States. After that time, the regular tv stations have a monopoly on telecasting the films.
2. During pay tv operations, the station may not carry any commercial announcements. Even promotion of other pay tv features is restricted to periods at the beginning and end of the pay tv program.

3. No serials or series programs may be presented on pay tv.
4. At least 10 percent of all pay tv hours (computed on a yearly basis, with no less than 5 percent in any given month) must be devoted to programing *other than* feature films and sports.
5. Stations engaging in pay tv programing must meet all requirements imposed on all other commercial broadcast television stations. They must also carry the following minimum amounts of nonpay tv programing:
   (a) During the first 36 months, no less than two hours daily in each of five days of the week, nor less than:
      12 hours a week for first 18 months,
      16 hours a week for 19 through 24 months,
      20 hours a week for 25 through 30 months,
      24 hours a week for 31 through 36 months.
   (b) After the 36th month, no less than two hours daily on each of the seven days of the week, nor less than 28 hours during any given week.

How and when did the FCC adopt rules for pay cable?*

In its Fourth Report and Order on over-the-air pay tv, the FCC recognized that a form of pay tv might also develop within CATV systems. It declined, at that time, to take any action or establish any rules with respect to such possible operations. The FCC was then opening up another proceeding (Docket 18397) in which it intended to investigate all aspects of origination by CATV systems. On October 10, 1969, the commission released its First Report and Order in that proceeding. In that action it authorized the commencement of pay cable operations without any of the restrictions it had imposed on over-the-air pay tv; stating that it did not desire to take any further action which would inhibit cable "unless and until experience [gives] some indication of a trend calling for action in the public interest."

The theatre owners and television operators objected vociferously to the establishment of a pay cable service that would be allowed to operate without any of the restrictions imposed on pay tv and insisted that the FCC reconsider its action. Accordingly, in a Memorandum Opinion and Order released July 1, 1970, the FCC acknowledged that pay cable would be "akin to subscription television and present . . . the same threat of syphoning programs away from free television." It imposed on pay cable a set of antisyphoning rules that were identical with those in force for the over-the-air subscription service.

---

* All CATV is, of course, a "pay" service. However, the term "pay cable" is used here to refer to a CATV service that provides certain specified programing for an additional fee.

In a different, but simultaneous, action that looked towards the leasing of CATV channels for various purposes, the FCC specified that "where channels are used for 'pay' programs . . . the restrictions specified in [the Fourth Report and Order] should be applicable including that no commercials may be carried on these channels."

What is the subsequent history of the regulation of pay cable?

From July 1970 to August 1971, the FCC was busy developing an entirely new regulatory scheme for CATV. In August 1971, Chairman Burch sent a letter to the Congress detailing the FCC's proposals in that area. The letter pointed out that there was a rule on the books prohibiting CATV systems from syphoning sports programs where a separate per-program or per-channel fee was charged. It stated that the FCC intended to require CATV systems to observe all legal sports "black outs," except where the system had contracted with the team owner to carry the games that were subject to the black out. The letter did not, however, consider any other aspects of pay tv by cable.

In early February of 1972, the FCC released a Report and Order promulgating new CATV rules to become effective March 31, 1972. Although the document stated that "it is [the FCC] that must make the decisions as to conditions to be imposed on the operation of pay cable channels," only one of the new rules (Sec. 76.225) deals specifically with that subject. Section 76.225 applies equally to "origination cablecasting" (programing presented by the CATV system itself) and to "access cablecasting" (programing presented by an enterpreneur who leases a channel on the system). It contains essentially the same antisyphoning and other program restrictions applied to over-the-air pay tv by Sec. 73.643(a) and (b) of the FCC rules. It does not include any of the rules regulating the relationships between the operators and program suppliers that are contained in Sec. 73.642. To this extent, the FCC has apparently retreated from its 1970 position. At that time, it announced that all of the provisions of the Fourth Report and Order would apply to pay tv by cable. The FCC was in favor of a new "wait and see" policy, under which detailed regulation in that area will not be imposed until after cable systems have had an opportunity to work things out by experimentation.

What is the latitude allowed to cable operators under the existing rules?

Under circumstances described above, considerable latitude is afforded the CATV operator. He can conduct the operation himself, or he can lease a channel(s), to third parties for that purpose. Such an operation can be conducted on any cable system, regardless of the size of the community or the availability of other service, so long as the system has sufficient capacity to provide the other services required by the rules. On systems with limited capacity, the Public Access, Educational Access, and Local Government Channels can all be used for pay tv (on either proprietary or lease basis) during periods when not required for their primary purpose. Unlike the situation with over-the-air pay tv, no special authorization must be obtained from the FCC to use these other channels.

If the cable operator elects to lease a channel to a pay tv entrepreneur rather than to conduct the pay tv operation himself, he must observe the provisions of Sec. 76.251(a)(ii) and (iii). This requires that leased channels be made available on a first-come nondiscriminatory basis and according to an "appropriate rate schedule." Conversely, the operator has no responsibility for (and is prohibited from controlling) the content of any program carried on leased channels, although he must maintain rules against transmission over those channels of obscene material, lottery information, etc.

Have the FCC rules with respect to pay cable now been finalized?

No. The FCC is currently investigating the possibility of relaxing the antisyphoning rules applicable to pay cable. This investigation is being conducted in a proceeding titled: In the Matter of: Amendment of Part 76, Subpart G, of the Commission's Rules and Regulations Pertaining to the Cablecasting of Programs for which a Per-program or Per-channel Charge is Made. The Docket number is 19554. In comments filed in that proceeding TheatreVisioN, Inc., run by Dore Schary, recommended that the two-year prohibition on feature films be changed to five years. This would give pay cable operators an additional three years to exhibit the films. Schary also recommended that all contractual provisions giving cable operators any exclusivity protection over broadcasters for the exhibition of films on either a regular or pay cable basis be outlawed. Conversely, all provisions giving television broadcasters exclusivity over any exhibition on cable would be prohibited as well.

What are the specifics that TheatreVisioN would like to see in the pay cable rules?

The TheatreVisioN group, and other proponents, believe that the FCC should adopt rules that will ensure a free and competitive market between broadcasters and pay cable operators. TheatreVisioN submits:

1. "Syphoning" of feature films will not be a problem if the broadcasters are free to buy and exhibit such products without regard to their purchase by, or exhibition on, cable.
2. For this reason, cable should be given no exclusivity or clearance protection over broadcasters with respect to feature films.
3. Conversely, broadcasters should not be allowed to obtain exclusivity or clearance protection over the exhibition of feature films on cable, whether such exhibition is on a regular or a subscription (i.e., per-performance fee) basis.

What are the conclusions of the pay cable proponents with respect to the existing rules on films available to pay cable?

Pay cable proponents feel that arguments of the theatre and broadcast interests against relaxation of the two-year rule for subscription cable are inconsistent with proper policy and are without merit.

They want the rule amended to allow the exhibition of feature film on subscription cable at all times, except between five and ten years following general release. TheatreVisioN, on behalf of the cable industry, states the following:

> Syphoning of feature film will be no problem if television is not subject to exclusivity or clearance protection in favor of cable. Conversely, subscription cable may well not survive if subjected to exclusivity or clearance protection in favor of television exhibition. The FCC will be doing less than its full job if it fails to consider, in the context of [Docket 19544], the problems created by this practice and the means of outlawing it.

What is the 1973 attitude of the Federal Communications Commission towards pay tv's future?

The FCC that, in 1968, adopted the Fourth Report and Order was a pro-pay tv body that was anxious to see pay tv implemented. Because there has been a substantial change in the membership of the FCC since that time, it is difficult to judge whether the present FCC, as a body, is strongly inclined towards pay tv. On the other hand, nothing has happened at the FCC level since 1968 to create the assumption that the present FCC would take an antipay tv stance.

What was the vote of the 1973 Federal Communications Commission on the 1968 Fourth Report and Order?

Only three of the seven men now with the Commission were on it in 1968. Commissioner R. E. Lee voted favorably; Commissioner Johnson concurred in the result. Commissioner H. Rex Lee had just taken his seat with the FCC at that time; therefore, he did not participate in the vote.

How does the FCC define a cable system?

The FCC, in Sec. 76.5(a) of their rules, defines a cable television system as:

> Any facility that, in whole or in part, receives directly, or indirectly over the air, and amplifies or otherwise modifies the signals transmitting programs broadcast by one or more television or radio stations and distributes such signals by wire or cable to subscribing members of the public who pay for such service, but such term shall not include (1) any such facility that serves fewer than 50 subscribers, or (2) any such facility that serves only the residents of one or more apartment dwellings under common ownership, control, or management, and commercial establishments located on the premises of such an apartment house.

What channels are usually leased for cable pay tv?

The midband channels (120 through 174 MHz), which are located between the fm (88 through 108 MHz) channels and the high-band television (174 through 216 MHz) channels 7 through 13, are leased for cable pay tv.

How many usable channels are there in midband?

There are nine usable channels in the frequency range between 108 and 174 MHz. These channels are called "A" through "I."

Why are these midband channels selected?

These channels perform the best with existing cable amplifier systems and equipment retrofitted for this purpose. New all-channel converters have 30 channels, which cover the midband and superband.

How are these nine superband channels provided by cable tv?

Nine additional channels use the superband frequencies (216 through 270 MHz) made possible by new amplifiers developed for this service.

Will pay tv use these new superband channels?

Midband channels are preferred. They are at lower frequencies and have less loss than superband channels.

What is the sub-band?

Sub-band frequencies are below channel 2 (54 MHz), which is a spectrum reserved by most cable television proponents for the two-way capability requirements of cable television service.

Will cable pay tv use sub-band frequencies?

Yes, for centralizing metering purposes. A return signal from some pay tv decoders will identify the program viewed by the subscribers for the monthly billing by the pay tv service.

What is the FCC's position on two-way capacity over a cable system?

The FCC requires that cable systems have the capacity for return communication on at least a nonvoice basis. Even rudimentary two-way communication, according to the FCC, can be useful in a number of ways: for surveys, marketing service, burgular-alarm devices, educational feedback, and others.

The FCC is not now requiring cable systems to install necessary return communication devices at each subscriber terminal. Such a requirement would be premature in this stage of the cable's evolution. For now, it is sufficient that each cable system be constructed with the potential of eventually providing return communication, without having to engage in time-consuming and costly system rebuilding. This requirement will be met if a new system is constructed either with the necessary auxiliary equipment (amplifiers and passive devices), or with equipment that could easily be altered to provide return service. When offered, activation of the return service must always be at the subscribers option. The capacity of a telephone pair, in proper distribution, many times will meet all requirements.

What is the time frame for cable systems providing two-way capability?

Existing cable systems must provide the capability by 1977. New systems being built from 1973 on must have the capability built into the system.

What is the importance of two-way capability to cable pay tv?

Some pay tv systems require this capability to provide centralized monthly billing for their planned subscriber service and cannot function in the cable television environment without this two-way feature.

What is the FCC policy with respect to expansion of capacity of a cable television system?

The FCC's basic goal is to encourage cable television use that will lead to constantly expanding channel capacity. Cable systems are, therefore, required to make additional bandwidth available as the demand arises. There are a number of ways to meet this general objective. The FCC uses the following formula to determine when a new channel must be made operational:

> Whenever all operational channels are in use during 80 percent of the time during any consecutive three-hour period for six weeks running, the system will then have six months in which to make a new channel available.

This requirement, according to FCC reasoning, should encourage use of the system, due to the knowledge that channel space will always be available. It should encourage the cable operator to expand and update his system continually.

Will most systems expand their capacity to provide pay service?

Yes. Pay tv is considered the most lucrative service which a cable television operator can provide. In fact, in areas where three network channels can be received by ordinary antennas, it is anticipated that pay tv will bring more subscribers to an existing cable television system than any other service.

What is the 1973 opinion of the President's Committee (Office of Telecommunications Policy) on pay tv?

The Nixon Committee's policy offers support for pay tv. They are, however, very strong in their opinion that viewers should not have to pay for popular, but expensive, programing that is already available on advertiser-sponsored television. The Committee is seeking the maximum use of broadband communications via cable. They realize that on cable's many channels there will not be sufficient advertiser or sponsor support available to produce all the programs that cable systems need to sell their services to the customer.

It is assumed, from evaluating the public statements, that the President's OTP realizes there must be a combination of subscriber-supported (pay tv) and advertiser-supported programs to fund the programing needed for cable television.

What are the technical system requirements needed for approval by the Federal Communications Commission for over-the-air service?

To be approved, pay tv transmission quality must be equal to that of existing color, or black and white transmission. This means there can be no apparent degradation created by the encoder. The decoder must be designed for insertion between the subscriber's antenna and television set. After the signal has been decoded, the performance must be comparable to that which the subscriber would have received from a program without decoding. Also, the insertion of this decoding device must make the overall subscriber reception no more susceptible to interference of any kind than the reception of conventional signals.

Are there any cable tv/pay tv restrictions written into cable operators' franchise agreements?

Yes. In most cities the theatre operators united and asked city councils to write a restriction against pay tv.

What does a cable operator do about this restriction?

There are only two actions that are necessary. First, he must notify the city council that the recent FCC decision invalidated all state, county, or municipal restrictions against pay tv. This was based on the grounds that the federal government has superceded the field and thereby preempted any local regulations prohibiting pay television. However, prior to commencing any pay tv operations in the community, the operator should meet with the local government to explain the nature of the proposed pay television service and what its advantages would be for the cable subscribers.

What has been the reaction of city councils to the cable operators offering this service?

Nearly all city councils realize that pay tv provides their constituency with another entertainment option; the scope of the city's existing entertainment options may determine their decision. In general, councils recognize that home and theatre atmospheres have changed and that the home subscribers are entitled to this service if they desire it.

Chapter 3

Economics of Pay TV

Who will prosper from pay tv?

Besides the pay tv proponents and the cable operators, pay tv will further the arts. With the tremendous return for the theatre groups from pay tv, the demand for new productions will grow. This will affect actors, writers, composers, producers, and others in the performing arts.

The philharmonic type of musical organization could prosper and make the field of music more desirable for America's younger set. Assuming only two million homes pay a fee of 50 cents to see a concert conducted by one of the famous maestros, a million-dollar return would result. This is far beyond the greatest dreams of any of America's present impresarios.

Currently, rock concerts are being recorded in stereo on video tape for uninterrupted play over pay tv as a means of attracting a younger audience to this new medium.

Consider the number of components which will be sold for the construction of 20 million decoders. Along with the components, replacement parts will probably be required. Those engaged in the manufacture of parts and subassemblies will gain a profitable market. This is another potential bonanza for manufacturers of cable systems, master antenna systems, and components for these engineered products.

As pay tv expands, so will the related industries of entertainment production and equipment manufacture. This diversified expansion will be profitable not only for private companies, but for the American economy in general.

Is pay tv a good business as a new venture?

It should be, if the economics of this chapter prove to be right. Pay tv is a new system, and a system is like the proverbial chain—no better than its weakest link. The software offered must be entertainment and educational matter that is desirable to the public. The hardware must be reliable, and the billing system must offer minimum burden to the subscriber and cable television operator.

What are the economics of cable pay tv?

Cable pay tv offers extra income to the cable operator and an opportunity to stimulate subscriber growth. It may be the only answer for making urban cable television a realizable service.

What are the economics of cable pay tv for the cable operator in an existing subscriber-saturated system?

In a saturated system, the cable operator has already provided for the wide-spectrum subscriber needs by installing a converter at his own expense; he is only seeking pay tv as a supplementary source of income. He will expect to receive income for leasing one, two, or three channels of midband or superband to a pay tv proponent. The cable operator in this environment looks upon pay tv as additional income. Naturally he would like to make minimum expenditures in support of pay tv, so that he can retain the maximum income. Most cable operators have indicated a preference for pay tv systems which do not require them to participate in the billing or promotional procedures. Nearly all of those operators who were interviewed said that they do not want pay tv collection responsibilities in addition to their monthly subscriber billing. In this respect, they favor the one-way systems in which the pay tv proponent has the total liability.

What are the economics for the cable television operator planning a new system in a fringe area?

Where the cable television service is needed to improve reception, pay tv is considered a plus service by the cable operator; it can accelerate subscriber growth. However, in this area cable television can usually be made profitable whether or not pay tv is an available service. Many of these areas are too small to justify pay tv service, unless it is interlinked via microwave to a master studio in a larger market area. A perfect example of such a system is the installation at Sarasota, Florida. TheatreVisioN, and Storer Broadcasting plan to interlink the 5 thousand subscribers in Venice, the 2 thousand subscribers in Englewood, and the 20 thousand subscribers in the Sarasota area. The plan is to link all of the systems in this territory (a two-hundred mile radius) to the master system operating from a major studio in Sarasota.

What are the economics for the cable operator planning a new system in an urban area?

In urban communities where an indoor antenna, or a simple outdoor antenna, will pick up three networks and a good independent vhf channel(s) the cable operator may be facing an economic fiasco. He must be able to motivate the home subscribers by offering better program fare via pay tv. The hardware in this area will not be as important as the software. The potential subscriber will have no interest in how he is billed, the features of scrambling or unscrambling, the simplicity of the operation of the decoder, etc. He will be motivated by the availability of first-run motion pictures, older classic features (Gone With the Wind, etc.), international sporting events, and local sports which he cannot see over advertiser-sponsored television. Actual survey experience has shown that all of the plus services (municipal channels, additional educational channels, home shopping, etc.) will not get the public to buy cable tv services.

The urban operator incurs a tremendous expense in cabling underground, gaining access to buildings, and installing superior converters which must meet rigorous performance standards established by the FCC. He will have a tremendous cash outlay; and it must be returned over a reasonable period of time (five to ten years) if he is to have a viable business. Cable television in the urban community is not a public utility and does not fill the necessities of life like power and gas. It is predicted by many experts that only good pay tv fare will develop the economics favorable to bringing urban cable television forth as a practical service.

What are the economics from the viewpoint of the pay tv proponent?

The pay tv proponent must pay approximately the following from the price of each subscriber ticket:

    50% to the film or sports producer
    10% to the cable operator
    20% for administration, promotion, and general management
    The balance of 20% is available to the pay tv proponent to write-off the cost of decoders and realize a reasonable profit.

Let us assume, for example, that a pay tv system has 20 thousand subscribers, each viewing 50 programs annually. At $2 per program ticket, each subscriber spends $100 per year. Thus, the system receives the combined total of $2 million in revenue annually. This amount would be divided as follows:

Programing	$1,000,000.00
Cable operator	200,000.00
Administration, promotion, and management ...	400,000.00
Equipment write-off and profit	400,000.00

Is pay tv economically sound for the smaller cable operator with a system potential of only 2000 subscribers?

No. It is not economically sound unless groups like Philco-Ford can adapt their low-cost studio system to the smaller cable television market. In that system the tickets are paid for in advance, and there is no complex or costly collection responsibility. A second possibility, as previously stated, is that the smaller community may join a regional network via microwave.

What are the economics of cable pay tv from the viewpoint of the subscriber?

Like all pay tv in the home, the family pays the price of one ticket and saves the costs of driving, parking, and babysitting. It is antici-

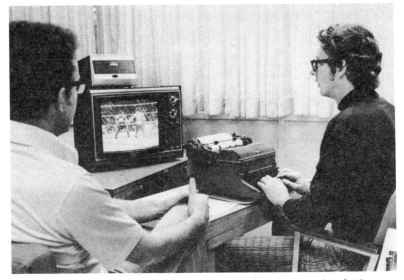

Fig. 3-1. City Editor Bill Zellmer and sports writer Jack Gurney covering the Foreman-Frazier fight via TheatreVisioN.

pated that the average family will save at least $5 by viewing at home; much more if the program is a sporting event. In Sarasota, the boxing match between Frazier and Foreman (Fig. 3-1) was seen by approximately 50 percent of the pay tv subscribers for a price of $3. The only alternative was to drive 45 miles to Tampa and view the fight on a theatre tv at $8 per ticket.

What is the FCC attitude on microwave services providing cable television systems with economical pay tv programing?

Sterling Manhattan, an affiliate of Time, Incorporated, has received favorable rulings from the FCC for interlinking cable television systems with special programing in the New York area. The FCC ruled against NATO (National Association of Theatre Owners) objections and considered program distribution via microwave to cable systems to be in the public interest. It is anticipated that unless the FCC continues to support interlinking systems, subscribers to smaller cable systems will be isolated from special entertainment fare. These interlinking systems presently include: Theta-Com's AML system for short, one-hop, multibeam microwave interlinkage; Laser Link's fm relay; or the joint development of Laser Link and Northrop Corporation (the Norlink system) for long-haul regional network service. It is expected that the FCC will rule favorably, since better programing in the home and freedom of choice is in the public interest.

Has anyone carefully analyzed over-the-air pay tv as a business?

Yes, Zenith Radio Corporation has. Based upon its six-year Hartford test and the analyses developed by Pay Television Corporation, they filed the facts in their 1969 FCC approval application.

What is the cost to a franchise holder going into business with Pay Television Corporation?

Here are a franchise holder's preoperation costs as Pay Television Corporation filed with the FCC in their 1969 application approved for Chicago:

The pay tv transmitting equipment (encoder) is to be furnished to the franchise holder in consideration of the franchise fee of five percent of subscription revenues. Accordingly, the $150 thousand capital cost of the encoder is excluded. The encoder installation cost, approximately $5 thousand, will be incurred by the tv station licensee and is included in the assignment application as part of his first-year operating costs. All other preoperation costs, as set forth below, are to be paid by franchise holder:

Decoder (includes cost of labor, supervision, and materials) $1,114,260.00

New subscriber costs (includes solicitation and installation of decoders after deduction of $10 installation fee) 1,515.00

Advertising and promotion 50,000.00

Shop repairs 3,000.00

Subscriber relations 1,750.00

Station time 15,000.00

General and administrative 166,955.00

Marketing (fixed expense) 72,400.00

Program (fixed expense) 17,100.00

Technical (fixed expense) 49,450.00

Total preoperation costs $1,491,430.00

What is Pay Television Corporation's estimated first-year cost of operation?

Operating Expenses:

Program direct costs .... $ 649,688.00

Advertising 116,944.00

Franchise fee 116,944.00

Bad debt provision 35,083.00

Decoder depreciation 423,274.00

Ticket production 71,130.00

Ticket reading 43,489.00

Cash processing 17,506.00

Subscriber relations 60,441.00

Service 54,176.00

Shop repairs 79,976.00

Satellite expense 4,607.00

New subscriber cost 8,060.00

Detachment costs 4,403.00

Station time 600,000.00

Lines and facilities 144,000.00

General and administrative 1,128,290.00

Marketing (fixed expense) 280,083.00

Program (fixed expense) 101,604.00

Technical (fixed expense) 187,104.00

Total expenses including depreciation $4,026,802.00

Less depreciation 444,342.00

Total cash flow operating expenses $3,582,460.00

Capital Expenditures:

Decoders (including cost of labor, supervision,
and materials) $4,811,310.00
Other technical equipment 43,488.00
Furniture and fixtures 155,000.00

Total capital expenditures $5,009,798.00

What is the total picture of the franchise holder fund requirements?

Preoperating Costs:

Pay tv transmitting .... None
Other preoperating costs .... $ 1,491,430.00

First-year Operations:

Operating expenses 3,582,460.00
Capital expenses 5,009,798.00

Total funds required $10,083,688.00

What are the sources of funds to be relied upon by the tv station licensee, franchise holder, and other participating persons to meet the costs of construction and operation stated above?

As stated above, the only fund requirements for the pay tv operation are those of franchise holder. Its sources of funds are as follows:

Source of Funds:

Estimated revenue from first year of operations . $ 2,338,875.00
Existing capital 7,744,813.00

Total funds available $10,083,688.00

What are the assumptions used in estimating cost of operations?

These are assumed to total 25 percent of the *gross* revenue, which also reflects actual experience with pay tv program procurement in both Hartford, Connecticut, and in California. In Hartford, program payments generally were based on 35 percent of *program* revenue. This 35 percent produced approximately the same amount as 25 percent of gross revenue would have produced, because decoder rentals (comprising 30 percent of gross revenue) were not included in program revenue.

In California, although as much as $33\frac{1}{3}$ percent of program revenue was paid for program products, actual payments never exceeded 25 percent of gross revenue, because decoder rentals were again excluded from the revenue base.

What are the decoder price and depreciation assumptions?

Zenith has estimated the cost of its new color decoder at $175, so straight-line depreciation during the estimated eight-year life becomes approximately $22 per year.

What are the station time cost assumptions?

Zenith is to perform the functions of tv station licensee and franchise holder. It was determined, for purposes of cost allocation, that the franchise holder division should be charged with a station time cost. This amount would be equal to what the franchise holder would pay if the tv station licensee were a completely unrelated entity. This cost was determined on the basis of the following considerations.

It was first assumed that independent uhf stations would be attracted by quantity time-buying arrangements. Such arrangements offer a greater revenue potential than that derived from affiliation with one of the three tv networks under standard rate practices (rates as published in *TV Factbook or Broadcasting Magazine*).

The average rate charged for Class "A" prime time by uhf stations was computed in market areas having five stations, as required by the FCC rules for STV. This average rate was used to determine an attractive rate for STV markets. In some instances, quotations by station representatives had to be utilized, because rates were not published in any broadcast publication. Thirty percent of this average uhf rate in each market was then assumed to be the STV rate for that market, just as the tv networks apply their 30 percent "rule of thumb" to individual station rates.

With respect to the Chicago market in 1969, application of the foregoing procedure produced the following results:

1. The average hourly rate for class "A" time was $887 per hour.
2. The "rule of thumb" (30 percent of line 1 rate) was $266 per hour.
3. The line 2 rate for 1500 hours per year was $399 thousand per year.

In order to increase the attractiveness of pay tv time sales, this $266 hourly rate was increased by 50 percent to an even $400 per hour ($600 thousand per year) for the first year.

What are the other cost assumptions?

All cost assumptions, other than those specifically mentioned before, are based on experience in the Hartford test, as related to new equipment and actual conditions that currently exist in the Chicago market. For example, payment of the Pay Television Corporation franchise fee is again assumed at the rate of 5 percent of net revenues. As percentages of projected income, all these other costs are relatively the same as those projected to the FCC in 1965 solely on the basis of the Hartford operation. Wages, equipment rental, and space for the planned Chicago operation will be substantially higher than in the Hartford test; these costs will be offset by the following efficiency factors not applicable to the Hartford operation:

1. The planned operation will use high-speed ticket production and electronic data-processing equipment for the administration of subscriber billings, collections, equipment, other inventories, and all other administrative activities.
2. The new types of color decoders require a simple attachment to the antenna terminals of the subscriber's receiver. This permits installations or detachments to be made by individuals with only rudimentary electronic knowledge and experience in only a few minutes time. This factor alone results in substantial savings in time and wage scales.
3. More efficient maintenance and repair operations, based on the modular design of the new decoders, will eliminate all servicing in the home. Instead, decoders will be replaced, returned to the service center (via satellite collection points, when required), and then processed and repaired on an assembly line basis by trained and highly skilled technicians. This permits facilities and personnel for electronic check and repair of decoding equipment to be combined into more efficient units.

What are the projected revenues to the franchise holder for first-year operations?

The estimated revenue to be derived by the franchise holder during the first year of operation is $2,338,875.00, computed in the following manner:

Fees paid by full-month subscribers, at estimated
    subscriber expenditure of $12.50 per month . . $2,432,500.00
Fees paid by new and terminating subscribers (for
    less than a full month) at estimated rate of
    $6.25 per month 166,250.00

Total subscription fees $2,598,750.00
Less volume discount of 10 percent 259,875.00

Net revenue $2,338,875.00

What are the assumptions as to program pricing and subscriber expenditures upon which the previous projections were made?

A survey of Chicago box office prices was used to estimate pay tv revenues in Chicago. Since the Hartford results were based on 1962 box office prices, it became essential to compare Chicago box office prices for 1962 with those for 1969. There was an increase from the 1962 Hartford prices to the 1969 prices in Chicago. The planned pay tv increase will be 35 percent, although the box office increase was higher. It can be assumed that Chicago subscribers would now be willing to spend $135 per year, since $100 per year was spent by Hartford subscribers based on 1962 box office prices.

Discounts based on quantity buying proved to be an effective marketing tool in Hartford. Therefore, it was determined that gross revenues per subscriber should be projected at $150 per year, with discounts for quantity buying averaging 10 percent, making a net revenue of $135 annually per subscriber.

On the basis of Hartford experience, Chicago prices for pay tv programs should be preferably less than the lowest-priced seat for the same type of entertainment outside the home. For instance, if theatres charge $1.25 for films showing before 6:00 P.M., a subscription film matinee should be $1.00; if films are $2.75 after 6:00 P.M., the pay tv charge should be $2.50. Opera, ballet, or concert tickets that cost $3.50 should be only $3.00 in the home. If the lowest-priced seat at a blacked-out sports event is $4.50, the pay tv price should be $4.00. Accordingly, competitive pay tv and box office prices are assumed to be as follows:

	Pay TV	Box Office
Motion pictures	$2.50	$2.75
Sports	4.00	4.50
Other	3.00	3.50

What are the assumed subscriber viewing patterns?

In order to test the $135 per year ($12.50 per month) assumed to be spent by the average subscriber, typical subscriber viewing patterns were constructed. This is based on the available 1500 hours of annual programming. With motion pictures and sports constituting the principal program categories, pay tv households will be principally oriented towards one or the other. A typical pattern of anticipated expenditure is projected below:

1. Film-Oriented Household
   42 of the 96 films available at $2.50 (average cost per viewer per film 83 cents) $105.00

6 of the 60 sporting events at $4.00 (average cost
per viewer per event $1.33) 24.00
7 of the 12 other types of entertainment at $3.00
(average cost per viewer $1.00) 21.00

Total price for 55 events $150.00
Less 10 percent discount 15.00

Total expenditure $135.00

If this same three-member household were to pay admission at the
box office for the same entertainment, the cost per year would be:

42 films at $2.75 ....$346.50
6 sporting events at $4.50 .... 81.00
7 other types of entertainment at $3.50 .... 73.50

Total cost for the same 55 events $501.00

During a one-year period, a sports-oriented household of three
persons could view the following events on the basis of approximately one pay tv program per week:

2. Sports-Oriented Households
28 of the 60 "blacked out" sporting events at $4.00
(average cost per viewer per event $1.33) $112.00
8 of the 96 films available at $2.50 (average cost per
viewer 83 cents) 20.00
6 of the 12 other events available at $3.00 (average
cost per viewer $1.00) 18.00

Total price for 42 events $150.00
Less 10 percent discount 15.00

Total expenditure $135.00

If this same household were to pay admission at the box office for the
same entertainment, it would cost the three-member family per year:

28 "blacked out" sporting events at $4.50 ....$378.00
8 motion pictures at $2.75 .... 66.00
6 other types of entertainment at $3.50 .... 63.00

Total price for same 42 events $507.00

What are the subscriber detachment assumptions?

During the Hartford experiment, the average monthly rate of detachment was 3.9 percent of total subscribers. However, the turnover rate
experienced by telephone companies, 1.5 percent per month, would
appear to provide a more realistic benchmark than the Hartford pay tv

experiment. It is believed that the Hartford rate can be cut in half through (1) better programing than was available under experimental conditions, (2) the use of color equipment instead of monochrome equipment, and (3) the use of a minimum program viewing charge, instead of the fixed decoder rental used in Hartford.

On the foregoing basis, it is assumed for Chicago projection purposes that (1) during the first year 1 percent of total subscribers at the end of any month will be detached during the following month; (2) this detachment rate will increase to 2 percent per month after the first-year novelty factor has worn off.

Why do over-the-air pay tv proponents believe that this service will be economically more viable now than cable pay tv?

Solomon Sagall, of Teleglobe Pay TV System, Inc., reports:

We are convinced that as far as this decade is concerned, the big revenue will come from over-the-air pay tv operations in the major markets.

Let's take as an example San Francisco-Oakland (where Teleglobe has granted an option on a franchise) with its 1.5 million tv homes. To wire that market by cable might take 10 to 15 years and a few hundred million dollars. In the case of over-the-air pay tv, the operation will break even with 25 thousand subscribers contributing, on a per-program basis, an average of $2.50 a week, and the decoders can be widely scattered.

With say, 100 thousand pay tv-installed homes, representing only 6⅔ percent of the 1.5 million homes paying $2.00 a month leasing charge for the decoder and only 60 percent of above, or 60 thousand subscribers contributing on the average $2.50 a week for programs of their choice, the gross income of the franchise holder will be $10.2 million.

According to *Broadcasting Magazine* of May 15, 1972, "CATV reaches now 6 million homes; the average system has 2,150 subscribers; and the largest, in San Diego, has 51 thousand connections." The cost of laying cables ranges from $4 thousand per mile in rural areas to more than $50 thousand per mile in large cities, and as much as $120 thousand a mile in New York City (*New York Times* of June 13, 1972, page 87). There are right now 28 over-the-air markets with over 500 thousand tv homes (ten of which have one million or more homes), nine markets with over 500 thousand homes, and there are fifteen more markets with over 300 thousand tv homes. Cable, compared with over-the-air, has the advantage of its multiplicity of channels and thus being able to offer simultaneously a choice of two or more pay tv programs. Realistically speaking, however, the provision of only one or two outstanding programs a week will tax the resources of program producers in the next few years. The single-channel capacity of an over-the-air tv station is, therefore, more than adequate for some years to come. A rule requiring cable operators to carry the scrambled programs in over-the-air pay tv markets has been proposed by the FCC.

Teleglobe, while actively engaged in the development of both over-the-air and cable pay tv technology, and while having an application pending for a CATV franchise in Queens, New York, is convinced that over-the-air pay tv is economically more viable and more lucrative than cable tv, at least during this decade. Our view is shared, among others, by Zenith and Pay Television Corporation (formerly Teco).

What economic problems will affect the growth of cable pay tv in urban and suburban markets?

The cost of the home subscriber terminal could have a serious affect on the cable operator's ability to fund and install a new system. The $100 to $125 terminal cost strains the cable operator's budget.

Are there any other developments on the horizon which could overcome these costs for the cable operator?

Dr. Kern H. Powers, in his report titled "Single-Cable Versus Dual-Cable Distribution Systems," presents an analysis which would replace the home subscriber terminal with a coaxial switch. Under these conditions, pay tv companies would furnish a one-way decoder. The two-way responsibility would be accomplished by a standard low-cost telephone line signaling device. If the pay tv company holds the title to the decoder, the cable operator will receive income with minimum home terminal equipment write-off.

Under what conditions can this principle be used?

Until television sets are better shielded, it cannot be offered in areas of direct pickup interference; the stations' signals penetrate the set directly, interfering with the signals on the cable. For example, in New York City, the idea is not practical due to the close proximity of the television stations' broadcasting antennas. For the adjacent Long Island and New Jersey communities, however, the principle becomes practical.

Has anyone used the dual-cable approach?

To quote Dr. Powers:

Several cable tv operators in the U.S. have already installed dual-cable systems with coaxial switches at the subscriber set to provide for more channels than can be handled on a single cable. In addition to the cost (about $4.00 per subscriber) of the coax switch, the cost of the distribution system is almost doubled. At least one CATV equipment manufacturer sells a line of equipment specifically designed for dual cable and by sharing power supplies and housing, the cost is substantially less than double that for single cable. However, most equipment manufacturers are now advertising distribution amplifiers designed to handle 24 to 30 channels on a single cable using midband and superband channels outside the normal vhf 12. A concurrent market has therefore developed for a line of set-top converters to tune these new frequencies. But because of the local oscillator radiation and image rejection problems in typical receivers, a host of problems must be solved by the cable operator in order to use these additional channels. Even if these problems are solved, the distribution equipments must have an increased dynamic range and a higher degree of linearity to handle the larger number of channels without an unacceptable harmonic and cross-modulation performance in a long cascade. Thus, these wide-band distribution equipments can be expected to cost more and the set-top converters must be supplied by the operator at a substantial increase in his investment cost per subscriber.

Has the Powers study gone far enough to make a real economic evaluation?

Yes, Powers explains that:

The development of both single- and dual-cable distribution systems have progressed to the stage that a realistic and objective cost comparison can now be made to determine what cost differential between these two system concepts can be expected to persist as a function of the subscriber density in a typical installation.

What are the assumptions in this analysis?

Dr. Powers assumes:

a module of a system consisting of one of the trunks fanning out from the head end over an area of five square miles. Subscriber density is a parameter ranging from 500 to 5000 per square mile typical of the suburban/urban market. The assumed specification is for delivery of 24 channels. For the dual-cable system the standard 12 vhf channels (with an upper frequency of 216 MHz) are applied in each cable; for the single-cable system we will use in addition eight midband channels and four superband channels extending to 240 MHz. The system layout takes into account the higher cable attenuation (at 240 MHz) as well as the 6-dB poorer cross-modulation performance (24 versus 12 channels) in the single-cable system. Accordingly, the amplifiers in the single-cable system are assumed to be operated at 3-dB lower gain and total output level (for similar cross-modulation specification) than those in the dual-cable system to provide equal picture quality in both distribution methods.

Degradations due to local oscillator radiation, images, and nonlinear distortions in the set-top converter are ignored, however, in the single-cable system.

Table 3-1 summarizes the results of the cost analysis, normalized to a per-subscriber investment value. It is seen that the "premium" for a dual-cable system is small to nonexistent in the suburban/urban markets, when the system layout is carefully normalized to comparable performance criteria. In fact, for higher subscriber densities (e.g.,

Table 3-1. CTV System Costs($/Subscriber) (Excluding Head End and Dead Run)

Item	500 Subscribers/ sq. mi.		5000 Subscribers/ sq. mi.	
	Single Cable	Dual Cable	Single Cable	Dual Cable
Amplifiers and power supplies	$ 75	$ 63	$ 14	$ 13
Trunk and feeder cable*	24	44	6	11
Passive components	4	8	3	6
Cable installation	51	54	14	16
Pole rearrangement	14	13	4	4
Pole rental (ten year)	25	22	7	6
Drop plus balun	7	14	3	6
Set-top converter	35	—	35	—
Coaxial switch	—	4	—	4
Total†	$235	$222	$86	$66

* The increased amplifier spacing permits a slight reduction in the total miles of cable for the dual-cable system.

† Does not compare $100, 1973 home terminals with coaxial switch, with telephone line return signaling which could have greater cost advantages for the dual-cable systems.

those represented by the top 50 U.S. markets where 65 percent of the tv homes are to be found) the cost per subscriber is actually *smaller* for the dual-cable system. In addition to the slight investment cost advantage, the dual-cable system offers the following bonus advantages:

1. Because of the higher permissible output level and lower cable attenuation, the dual-cable system can serve a 70-percent greater area per trunk for a given performance level, than the wideband single-cable system.
2. It has higher system reliability from the coaxial switch over the set-top converter.
3. It has higher initial installed capacity (upgradable to more channels without the laying of additional cables).
4. The midband capacity for internal system use and for leased communication services on channels not readily accessible to the normal subscribers is available in the system.

The following assumptions are also implicit in the analysis:

1. The systems are constructed for 100-percent penetration of the service area.
2. Every trunk/bridging amplifier in the distribution area feeds subscribers.
3. Each output from the trunk/bridgers feeds two distribution amplifiers (line extenders) in cascade per feeder line.
4. Alternate trunk amplifiers contain agc.
5. A minimum signal level of 3 dBmV (75 ohms) is supplied to each subscriber, and a minimum isolation of 9 dB is maintained at the subscriber terminal that is the farthest distance from the trunk/bridger.
6. One cable power supply unit serves three amplifiers.

Table 3-2. Specifications

	Single	Dual
Number channels/cable	24	12
Bandwidth	240 HMz	216 MHz
S/N (at end of cascade)	43 dB	43 dB
Cross-modulation (at end of cascade)	—51 dB	—51 dB
Noise figure	12 dB	12 dB
Gain/amplifier	17 dB	20 dB
Amplifier spacing (trunk)	2100 ft.*	2700 ft.*

*New high-quality "Dynafoam" cable is assumed. This permits up to 35-percent increase in cable spacing over older cable types.

Table 3-3. Assumed Unit Costs (Common)

Item	Cost
Trunk cable	$300/1000 ft.
Feeder cable	140/1000 ft.
Drop cable	38/1000 ft.
Power supply	$212.00
4-way tap	$ 11.50
2-way splitter	17.50
Line termination	1.05
Pole rearrangement	$500/mi.
Pole rental (10 year)	$880/mi.

Table 3-2 gives the assumed system specifications, while Table 3-3 and Table 3-4 give the assumed unit costs, derived from prices of typical American equipment meeting the assumed specifications. The cost of certain common items (e.g., head-end equipment) are omitted.

Table 3-4. Assumed Unit Costs (Uncommon)

Item	Single Cable	Dual Cable
Trunk amplifier	$314	$470
Trunk amplifier (agc)	365	560
Trunk/bridger	470	880
Trunk/bridger (agc)	515	970
Distribution amplifier	290	410
Cable installation	$1800/mi.	$2100/mi.
Set-top converter	$35.00*	—
Coaxial switch	—	$3.50†

* $35.00 is 1973 minimum cost per unit for reasonable production runs of quality converters.
† May be low for 1973.

What is the purpose of Dr. Powers' report to the industry?

The purpose of Dr. Powers' report is to show that dual-cable systems are not much more expensive than single. This report is also designed to stimulate re-evaluation of the need for converters in home subscriber terminals; converters are a possible source of trouble and added cost in getting new installations started. Also, if the converter section is taken out of the pay tv decoder, a further savings can be made by this new industry since the encoding can be done "on-channel" and decoded without changing the channel frequencies selected for the pay tv service.

What action, by pay tv proponents, could accelerate the acceptance of this dual-cable idea?

The dual-cable idea could be more readily accepted if the proponents furnished the cable operator, at a low cost, with the accessory coaxial switch and return path circuit. Acceptance will also be accelerated if someone markets an inexpensive response terminal that complies with the minimum FCC requirements and is compatible with one-way decoder systems (Home Theater Network, TheatreVisioN, etc.).

Chapter 4

Over-the-Air Pay TV

PAY TELEVISION CORPORATION

What is Pay Television Corporation?

Pay Television Corporation was organized in 1949 when it was spun off to Zenith Radio Corporation's shareholders. The company presently owns, under certain conditions, all the U.S. patent claims obtained by Zenith for encoders and other equipment used in pay tv transmission. Pay Television Corporation is, therefore, responsible for introducing commercially the Zenith system of subscription television; Zenith will act only as a manufacturer of encoders and decoders.

Pay Television Corporation is managed by Pieter E. van Beek, president, who served for twenty years as an executive of Zenith. George F. Wiemann, executive vice president and treasurer, was formerly a vice president and controller of Twentieth Century-Fox Film Corporation.

The company intends to establish and own subscription television operating companies in some major markets and to franchise pay tv systems in other markets.

Does Pay Television Corporation have FCC approval for its system? If so, when was it received?

The Zenith system, used exclusively by the Pay Television Corporation for over-the-air pay tv, was granted technical approval by the FCC on August 25, 1970.

What is the nature of the Pay Television Corporation service?

Zenith's system for over-the-air pay tv uses scrambling and unscrambling techniques, making certain quality broadcast programs available only to those who select and pay for them. The system has three basic elements: (1) an encoder for transmission, (2) a decoder for reception, and (3) a subscriber ticket for billing. The encoder, located at the local pay tv station, scrambles both the sound and picture signals before they are broadcast (Fig. 4-1). The decoder is a compact unit furnished to subscribers. It is quickly and simply connected to the antenna terminals of any home television set, regardless of make or model. When activated, the decoder restores the scrambled picture and sound signals to their original form; in color or black and white, and on vhf or uhf.

Fig. 4-1. Pay tv is received in its coded form (left) in Zenith's (Pay Television Corporation) over-the-air Phonevision system. The decoder on the television set (right) clears up the picture and sound.

To watch a program, the subscriber inserts a ticket in the decoder (Fig. 4-2). The ticket identifies the user; the decoder unscrambles the sound and picture for the duration of the program selected. The purchase of each program selected is recorded on the ticket to provide a basis for billing through a centralized computer system.

How does the encoder work?

The encoder is utilized in the transmission of subscription television programs. It consists of electrical equipment designed to scramble audio and video broadcast signals in order to render them unintelligible. The encoder is normally located at the business premises of a local Pay Television Corporation franchised operator. It is connected

Fig. 4-2. This decoder, developed by Zenith Radio Corporation, is part of the Phonevision over-the-air system approved by the FCC.

by coaxial cable with one or more local television stations. Any such station can then route its subscription television program signals through an encoder before feeding them through its transmitter. All receivers within the coverage area of such subscription broadcasts will then reproduce the scrambled picture and unintelligible sound described above. The encoder also generates additional broadcast signals called "air code" that permit subscribers to decode the scrambled signals and to restore the normal picture and sound.

How does the decoder work?

The decoder is an instrument furnished to any subscriber desiring the service. It is simply connected between his antenna and the antenna terminals of his television receiver. Upon instruction from the transmitted air code, the decoder can restore the scrambled video and audio signals to their original form. In addition, the decoder records each program purchase and provides the basis for subsequent billing. The decoder has four controls on its front panel. To the far right is a channel selection knob, which must be properly set for the channel of the television station broadcasting the subscription program. To the left of this knob are three additional controls. The center one acts as an on-off switch. When turned to the "on" position, this center knob connects the decoder to the receiver.

The two knobs on either side of the on-off switch are program-setting controls. One knob has eight dial positions, numbered "1" through "8"; the other has 12 dial positions lettered "A" through "M." Program code information is given to the subscriber in a form such as "6B," which means that one program code setting knob must be dialed to the "6" position and the other to the "B" position.

On the top of the decoder is an opening covered by a slide. The slide conceals access to a slot into which the subscriber periodically inserts a ticket. The decoder cannot function without the ticket in position in this slot.

What is the subscriber ticket?

Technically speaking, the system requires the subscriber to "prepare" the decoder so that various electrical signals, generated by the encoder and decoder, can perform their preassigned decoding functions. In order to accomplish this, the subscriber is periodically issued a ticket, which he must insert into the decoder. This ticket serves three basic functions: (1) it identifies the subscriber; (2) it records the purchase of each subscription program selected, providing a basis for subsequent billing; and (3) it contains some circuitry which, once it is inserted into the decoder, assists in accomplishing decoding.

What stations have applied to the FCC for approval to use the Pay Television Corporation's service?

Pay Television Corporation has agreed to purchase Los Angeles uhf Channel 52 from Kaiser Broadcasting Corporation. This purchase is subject to FCC approval. The company is also negotiating the purchase of tv stations, for subscription television purposes, in other major markets.

When does Pay Television Corporation plan to operate as a pay tv service?

The Zenith system of pay tv was extensively tested, both commercially and technically, in a six-year experiment in Hartford, Connecticut. On the basis of this experience, detailed planning for large-scale commercial operation has been completed. Pay Television Corporation anticipates that pay tv service could start in the early part of 1974.

How does the Pay Television Corporation (Zenith) concept vary from other proposed over-the-air pay tv services?

The system is patented and is a total proprietary development with encoding, decoding, and computing techniques exclusive to the Phonevision system (Fig. 4-3). In Zenith's encoding process, the video is scrambled by cutting the tv picture into 35 horizontal segments of seven lines each and continuously oscillating alternate segments. Concurrently, the position of the divisions between segments are shifted randomly, creating a visual effect of vertical movement. Simultaneously, the system has polarity inversion of black and white video signals so that the scrambled picture has a negative appearance. The sound is made unintelligible by an upward shift of the audio frequencies transmitted. Zenith's decoding process reconstitutes the picture and returns it back to a positive polarity transmission. The audio decoding is achieved by shifting the incoming sound to its original frequency.

To avoid cheating, there is a proprietary security system which changes the code from program to program. The code change is under the control of the station transmitting the program.

How does the viewer know what programs are being offered over the pay tv channel?

The subscriber periodically receives a program guide, listing each subscription program to be offered during the period. Also listed are the date and time of broadcast, program price and code number, and the channel number of the television station broadcasting each program.

The subscriber selects the program he wishes to see at a designated time and price. He dials the program code number, dials the proper channel number, and turns the decoder on-off switch to the "on" position.

How is the subscriber billed?

Shortly before the end of the current month or validity period, the subscriber is sent his next ticket and a self-addressed envelope for returning the previous ticket. The subscriber cannot continue to use the old ticket beyond its intended use period. The coding of programs in the succeeding period is established in relation to a ticket having a different matrix configuration.

If the subscriber fails to return his ticket, he will not receive a subsequent ticket and is thereby prevented from using his decoder. When the subscriber's ticket is received at the operating company it is

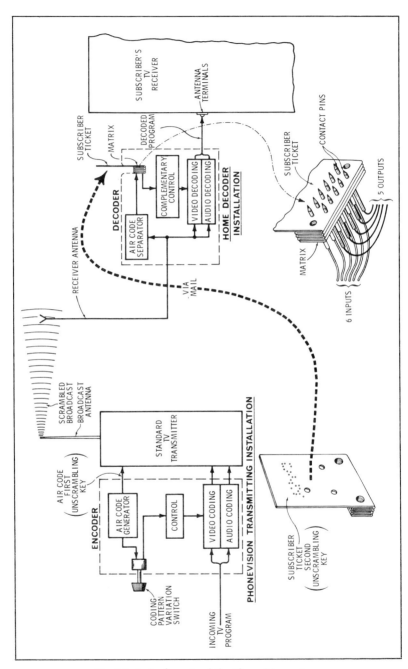

Fig. 4-3. Basic elements of the Phonevision system.

briefly examined for damage or mutilation. It is then fed into an instrument that die punches the identification and program use of the ticket. When fed information received from the ticket, a computer activates a billing machine which produces an invoice for the subscriber. This invoice shows the identity and price of each program viewed, the decoder rental charge, and the total amount owed. Information regarding this latter amount is also stored in the computer's memory. The bill is then mailed to the subscriber. When the subscriber's payment is received, an IBM card is prepared and the payment information is again stored in the computer. If the payment has not been received when the subsequent invoice is being prepared, the computer will automatically add the previous balance to this subsequent invoice. Similarly, previous payments will show on any future invoice.

Subscriber tickets received at the operating company are processed into invoices on an almost completely automated basis. This system can efficiently accommodate a very high volume of business. None of the equipment required in this processing is unduly expensive or complicated; even a relatively small Phonevision operation can afford this type of processing. Furthermore, small Phonevision operations could combine their processing at one central processing center in order to obtain the volume necessary for efficient utilization of the equipment.

What programs are planned for Pay Television Corporation's system?

Pay Television Corporation plans to provide the best in films, Broadway plays, operas, ballets, concerts, sports "blacked out" locally, and special events not shown on local television.

TELEGLOBE PAY-TV SYSTEM, INCORPORATED

What is Teleglobe Pay-TV System?

Teleglobe, one of the pioneering companies in the field of subscription television, was formed in 1957. It has been engaged, since its inception, in the development of pay tv technology for use over-the-air and over cable. Teleglobe was one of the chief proponents in the long battle before the FCC and Congress for authorization of pay tv (STV) on a permanent nationwide basis.

The company has been granted nine patents in the United States and several in Canada and now has several pending patent applications. Teleglobe Pay-TV System is under the leadership of tv pioneer Solomon Sagall.

Does Teleglobe have FCC approval for its system?

The Teleglobe Pay-TV '410' System was approved by the FCC on March 1, 1973.

What is the nature of the Teleglobe Pay-TV System, and how does it vary from other proposed systems?

The Teleglobe Pay-TV '410' System employs sophisticated concepts which are subject to pending patent applications (Fig. 4-4). Accordingly, the operation of the system, at the transmitting station and at the subscriber's home, is described in general terms.

At the station: Encoder equipment is connected to the standard visual and aural tv transmitters, STV programs are encoded, and each program is individually coded. The coding provides a high degree of security. Both the video and audio are scrambled. The Teleglobe Pay-TV '410' System works effectively with either monochrome or color television programs.

At the subscriber's home: The Teleglobe franchise holder furnishes the subscriber with an "Electronic Home Box Office," which consists of a decoder and an accounting mechanism (Fig. 4-5). It performs the dual function of recording information for billing and decoding the subscription program. The Electronic Home Box Office is connected externally to the antenna terminals of the tv set. This can be completed in a few minutes, using only a screwdriver. The back of the tv set does not have to be removed; consequently, the franchise operator will not have any responsibility for servicing the subscriber's tv set.

Billing methods: Teleglobe pioneered the concept of centralized metering by means of a physical link between the central office of the pay tv operator and the subscriber's home. The coaxial cable provides such a link in the case of CATV systems. Low-grade signal circuit wires, as provided by the telephone company for data transmission, may be used for metering information for the over-the-air pay tv. Teleglobe favors centralized metering as the ultimate method to be introduced in areas of high subscriber concentration. Initially, Teleglobe is employing an alternate method of individual decoding cards.

Teleglobe, in the design of its system, has been aiming at the maximum possible degree of security. They also hope to achieve the greatest possible simplicity in operation and, thus, the least burden on the subscriber.

A monthly or weekly pay tv program guide will be supplied to the subscriber. This information would also be carried in the tv section of the local newspapers. To receive a program, the subscriber performs the following simple operations:

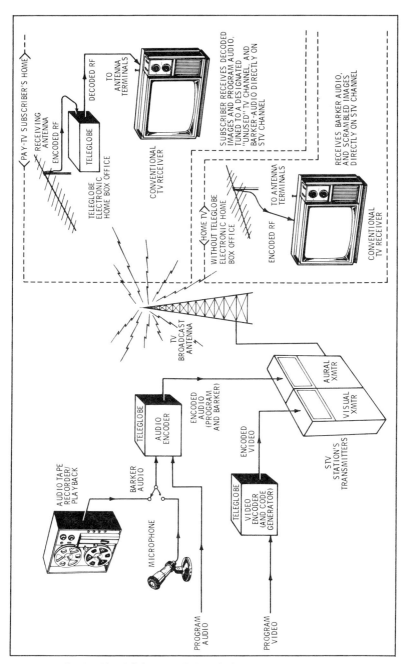

Fig. 4-4. Pictorial diagram of the Teleglobe Pay-TV '410' System.

1. Turn on the television set at the appropriate time and set it to the pay tv viewing channel (an otherwise unused and unassigned tv channel in the area).
2. Insert the billing keycard into the Electronic Home Box Office unit.
3. Set the unit and card for the program to be subscribed to, and activate the unit. (The program subscribed to is recorded on the keycard.)

This billing unit will be associated with the system's coding, which will be changed on a program-to-program basis. The code will be established by switch settings at the transmitter, corresponding to connections made by the keycard and/or the switches at the decoder unit. These will be altered before each pay tv program. A single keycard will cover several pay tv programs (e.g., as many as 30 pay tv programs in a period).

Fig. 4-5. The Teleglobe Model 410 Electronic Home Box Office consists of the decoder and the accounting mechanism.

A single card may include the code connections for several programs. At other times, a single set of code connections may be provided by the card, with additional code connections being made by switching the set for each individual program.

For decoding and billing purposes, the keycard will be inserted for each program, and program usage will be recorded on the keycard.

The decoder unit will be capable of being switched in such a manner that a nonsubscription broadcast may be viewed during a subscription broadcast. When the subscriber switches back to resume viewing the pay tv broadcast, he will not incur any additional charge.

A new card will be sent to the subscriber a short time before each new series of subscription programs. If a subscriber does not maintain payment for the service, the next keycard will be denied.

By providing a number of code burst tones an adequate number of variations are possible for program codes. Several cards or sets of programs will be provided before any program codes need be repeated. Therefore, it could be several months before a program code on a given card could be successfully used again for decoding.

To prevent code connections from being made by cards other than the Teleglobe card, additional connections are made via the card itself and are not otherwise accessible. To prevent others from duplicating the card, and surreptitiously providing it to others, the card is complex and economically feasible to produce only in large quantities. Only large established "counterfeit" operations would have the capital to reproduce the cards and these operations could be regulated by law enforcement. Copyrighting the cards or prosecution for patent infringements would also make such "pirate" operations uneconomical.

What stations have applied to the FCC for authorization to use the Teleglobe System?

Lincoln Television, holder of a construction permit for KTSF (Channel 26) in San Francisco, has applied to the FCC for authorization to engage in subscription television by using the Teleglobe Pay-TV '410' System.

When does Teleglobe plan to operate as a pay tv service?

Teleglobe commercial policy is to grant franchises for the use of its pay tv system to third parties, as well as to engage directly in pay tv operations in selected markets.

Telease of Milwaukee has obtained from Teleglobe an exclusive license to use its pay tv system in several major markets. An application is in the process of being filed with the FCC for a construction permit in respect to Channel 24 in Milwaukee. Telease is also in advanced negotiations for engaging in over-the-air pay tv in Washington, D.C. and in Los Angeles.

Teleglobe plans for direct pay tv operations in selected major markets will be announced in due course.

How does the viewer know what programs are being offered over the pay tv channel?

This information is received by means of the barker audio, from schedules in local newspapers, and from the operating company in weekly or monthly pay tv guide.

What programs are planned for the Teleglobe System?

The programing outlook is to provide better programs free of annoying commercials. Solomon Sagall, President, states:

> With the exception perhaps of a few programs, no viewer will want to pay for the present-day tv fare. Pay tv will have to provide outstandingly superior and challenging material to be able to induce viewers to pay. The quality of programs is crucial for success of pay tv.
>
> The problem that is facing pay tv operators is how to obtain programs of satisfactory nature during the initial stages of the pay tv service, when the number of subscribers is relatively small. Indeed, it is a vicious circle: one needs good programs to attract subscribers, one needs subscribers to be able to purchase good programs.
>
> Teleglobe will actively sponsor the production of pay tv programs as well as encourage the formation of new production companies.

BLONDER-TONGUE LABORATORIES, INCORPORATED

What is Blonder-Tongue Laboratories, Incorporated?

Blonder-Tongue Laboratories is a New Jersey corporation led by inventors Isaac Blonder and Ben H. Tongue. The company has achieved national prominence as the largest designer and manufacturer of custom master antenna television systems in the United States. Its total product line has in excess of four hundred items for the MATV, CATV, and home television markets.

Blonder-Tongue has three interrelated corporations to provide pay tv service. Blonder-Tongue Broadcasting Corporation holds a tv station license from the FCC, operating transmitter and studio programing 28 free hours weekly. It is supported by advertisers. BTVision operates the pay tv system as authorized by the FCC and also promotes subscription television. BTVision collects and disburses decoder rental and program income. Blonder-Tongue Laboratories licenses use of BTV systems to BTVision and builds and leases decoders.

Does Blonder-Tongue have FCC approval for its system?

Yes, equipment responsive to the FCC's specifications was developed in 1971.

What is the nature of the Blonder-Tongue pay tv service?

The BTVision encoder suppresses the horizontal and alters the vertical synchronizing pulses, causing the received picture to have a continuous random horizontal tear with a 10-Hz vertical oscillation. When scrambled, the picture content is totally obliterated. The program sound is placed on a subcarrier, leaving the regular sound channel (barker) available for announcements to the potential program purchaser. At the customer's home, the received signal is fed to a BTVision decoder (Fig. 4-6), which is activated by a single button that restores the original signal with full quality. The barker audio can be heard at all times from a small speaker located in the decoder, even when the tv set is not operating. With the BTVision system, each time the decoder button is pushed, a "real-time" ticket is generated by an integral strip printer. This ticket, marked with the program identification number and price, is stored in the decoder until mailed in with the monthly payment.

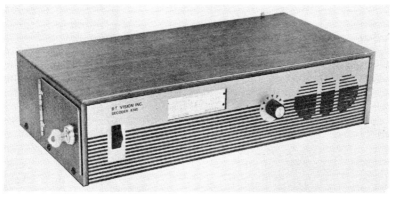

Fig. 4-6. The Blonder-Tongue decoder has a self-contained speaker which provides the "barker" audio.

What stations have applied to the FCC for approval to use the Blonder-Tongue service?

FCC approval for over-the-air pay tv transmission to the Greater New York and New Jersey metropolitan area was granted to the Blonder-Tongue Broadcasting corporation. A construction permit for the tv station (Channel 68) was simultaneously approved. The station was the first pay tv facility to be licensed by the FCC in the United States and will use the BTVision system for over-the-air transmission of scrambled television programs.

When does Blonder-Tongue plan to operate as a pay tv service?

Blonder-Tongue plans to begin pay tv service in 1974.

How does Blonder-Tongue's concept vary from other proposed over-the-air pay tv services?

1. Barker available to promote programs during scrambled transmission
2. Real-time ticket printer
3. Horizontal and vertical scramble

How will the viewer know what programs are being offered over the pay tv channel?

The subscriber will tune the pay tv channel and listen to the sound barker. The barker will describe the program, the time it can be seen, and the charge per program.

What programs are planned for the Blonder-Tongue systems?

Current motion pictures, locally blacked out sports, hit musicals, and championship boxing matches are examples of possible BTVision programs.

What is over-the-air pay tv?

Over-the-air pay tv is now authorized by the FCC for transmission, by qualified television stations, of motion pictures, legitimate stage plays, sports events, and other attractions to be received on conventional television sets, on a per-program basis. Rules and regulations for this service, and the areas where the FCC will authorize the service, were given in Chapter 2.

Who are the over-the-air pay tv proponents?

The over-the-air pay tv proponents are listed here, according to the length of time that they have been proponents. The first three systems have gained approval by 1973.

Pay Television Corporation, formerly TECO, is the current representative of the Zenith Radio Corporation's interest in pay tv.

Teleglobe Pay TV System is led by Solomon Sagall. He has had periodic support from Bartell Media, Cinerama, Durwood Theatres, and others.

Blonder-Tongue, an equipment manufacturer guided by inventors Ike Blonder and Ben Tongue, has fostered the BTVision concept for many years.

TheatreVisioN is the latest proponent, having obtained licenses from Laser Link Corporation in 1972. TheatreVisioN, known as TVN, is run by motion picture executive Dore Schary.

THEATREVISION, INCORPORATED

How does TheatreVisioN's over-the-air system differ from its cable pay tv system?

Over-the-air pay tv may be transmitted on only one television station in one market, under the rules governing such transmissions promulgated by the FCC. Since cable television systems can transmit as many as thirty or more television programs, there are no limits on the number of pay tv programs a cable system could carry in any market area. The physical appearance and function of Theatre-VisioN's home terminal unit is the same for cable and over-the-air pay tv.

What is the TheatreVisioN over-the-air pay tv system?

TheatreVisioN has developed a proprietary method of scrambling television images that completely distorts the sight and sound of the televised program (Fig. 4-7). Patents covering the TheatreVisioN technology have been issued, and the system is to be submitted to the FCC for type acceptance in the near future.

How does the TheatreVisioN over-the-air system function?

TheatreVisioN has developed a method whereby it can interject into its encoding equipment certain proprietary circuitry that can scramble both video and audio of a television program. The encoder is installed at a conventional television broadcast station authorized to televise

Coded (Scrambled) Image   Decoded Picture

Fig. 4-7. TheatreVisioN over-the-air system.

pay tv programing; it feeds pulses into the transmitting equipment that will scramble certain programs. At the consumer's television receiver, decoding devices attach externally to the television set and reconstruct the original image, when activated by a plastic electronic ticket (Fig. 4-8). This ticket is another proprietary development of Theatre-VisioN. The broadcaster is able to offer a wide variety of programing not available on conventional television for which he receives a charge per program.

How does the TheatreVisioN over-the-air system differ from other airlink pay tv methods?

1. The viewer uses a plastic electronic ticket to pay in advance for the program he wishes to see on TheatreVisioN. Some other proposed over-the-air pay tv systems require either monthly billings or flat fee charges, which a viewer must pay whether he watches the programing or not.
2. Other proposed systems require the viewer to set decoding dials before viewing a program, and they are billed afterward.

What is the market potential for over-the-air pay tv?

Under the rules of the FCC, only markets with five or more television stations presently operating will qualify. Only one of the five stations may qualify for pay tv transmission over the air. This limits over-the-air pay tv to about twenty of the one hundred major markets in the United States. There are, however, many smaller markets with five or more stations that would qualify for this type of service.

Fig. 4-8. TheatreVisioN's plastic ticket, with coded program indentification, is fed into the over-the-air decoder to unscramble the encoded program.

How does a viewer obtain tickets for the TheatreVisioN over-the-air method of transmission?

1. In each of the major markets authorized for pay tv, TheatreVisioN will establish a local outlet. It will feed motion pictures and other entertainment through the outlet by direct link to an authorized television broadcast station.

2. The organization will mail or sell tickets to consumers having TheatreVisioN decoders (which it will install), and thus provide a direct tie with home viewers.
3. The tv station carrying TheatreVisioN will either sell time on the air to TheatreVisioN for those special transmissions or will share a percentage of the income from such programing.

How soon will TheatreVisioN's over-the-air system become available?

TheatreVisioN's system prototype will be bench tested in 1973. It should be submitted to the FCC by the end of 1973 to obtain required approvals needed to become a valid service.

Are any other companies developing over-the-air systems?

Yes. An organization called Vue-Metrics, run by Attorney Seymour M. Chase, plans to have a system in 1973.

What over-the-air pay tv applications were filed with the FCC?

As of the spring of 1973, the following applications were reported.

Boston:	Boston Heritage Broadcasting, WQTV, Channel 68
Chicago:	WCFL-TV, Channel 38, is now being transferred to Pay Television Corporation
Detroit:	WJMY-TV, Channel 20
Newark:	Blonder-Tongue Broadcasting was granted a construction permit for WBTB-TV, Channel 68, July 26, 1972.
Philadelphia:	Vue-Metrics is now applying for vacant Channel 57. (The original application was for Channel 23.)
Philadelphia:	Radio Broadcasting Company is applying for Channel 57.
San Francisco:	Lnicoln Television, KTSF-TV, Channel 26
Sacramento:	Applications are in process.
Los Angeles:	Applications are in process.
Milwaukee:	Applications are in process.

Chapter 5

Hotel/Motel
Pay TV Systems

Why has hotel pay tv started before pay tv over cable systems?

These systems were able to start without concern for the complex rules of cable television and the decisions which were rendered by the FCC in 1972.

Are hotel/motel pay tv systems completely free of FCC rules?

No, those systems which have a studio in one hotel and microwave to other hotels in the area must obtain FCC approval.

How large is the hotel/motel market for pay tv?

Every hotel room is a potential market for pay tv. Three different types of pay tv are used in this market: (1) cartridge players, which require no centralized studio; (2) small systems which adapt to MATV; and (3) large systems having one major studio that serves hotels and motels via microwave.

Has any organization proven that hotel/motel pay tv systems will meet with success?

Yes, the Atlanta test of Trans-World Communications' Tele/Theatre has been termed a success.

Is Tele/Theatre planning to expand its service as a result of these tests?

Yes, they have announced plans to show current motion pictures on closed-circuit television in hotels in more than 25 major cities.

How long was the Atlanta test conducted?

The test results were reported during a year of service. On December 1, 1971, Trans-World Communications (the closed-circuit television division of Columbia Pictures Industries) began programing current motion pictures into the 1000 rooms of the Hyatt Regency Atlanta. Jerome S. Hyams, senior executive vice president of Columbia, termed the Atlanta test an "unqualified success."

Underscoring this evaluation, Hyams announced plans to extend Tele/Theatre into six more hotels in the Atlanta area. Also, the company revealed that it is presently installing its Tele/Theatre equipment in New York and Toronto hotels. They plan to initiate programing in those cities by the latter part of this summer. Chicago, Miami, Honolulu, Houston, and San Francisco will follow shortly. Hyams reported that their plans call for Tele/Theatre to be available in approximately 29 thousand hotel rooms in six cities by December 1972; these figures would increase to approximately 160 thousand rooms in 25 cities by December 1973, representing a potential viewing audience in excess of 43 million persons annually.

William J. Butters, vice president and general manager of Trans-World Communications, summarized the results of the pay tv service at the Hyatt Regency Atlanta:

> During the first three months of operation we have achieved successively greater acceptance by hotel guests of our Tele/Theatre service. The average daily requests for movies have more than doubled from our initial experimental stage in December, to the last full month of operation (February 1972). We now are at a point where the Tele/Theatre service has been utilized by nearly one fourth of the guest registrations at the Hyatt Regency Atlanta.

	Dec.	Jan.	Feb.
Percent of Guest Registrations	14.2%	22.3%	24.4%

One of the most significant findings that has emerged from our operations to date confirms that we are able to attract that segment of the public that has lost the movie-going habit. Tele/Theatre represents an effective extension of the theatrical box office. Our findings indicate that a full two thirds of the viewing of Tele/Theatre at the Hyatt Regency Atlanta occurs during the nonevening hours, including early morning and late night, when most movie houses are not open. Only one third of the total viewing of Tele/Theatre occurs during the heart of the evening.

Daytime 9 A.M. to 7 P.M. 31%	Late Night 11 P.M. to 9 A.M. 36%	Total Nonevening 67%

Evening
7 P.M. to 11 P.M.
33%

Tele/Theatre is obviously a convenient way for people to see movies. For a variety of reasons, many people, especially those over 30, have lost the movie-going habit. Tele/Theatre is getting them back to the movies simply by bringing the movies to them.

What motion pictures were shown to the hotel guests during these tests?

Only the best motion pictures available at the time. During this test period in Atlanta, Tele/Theatre offered the Hyatt Regency Atlanta guests a variety of motion pictures from five different distributors. A fee of $3 was added to the guest's hotel bill for each film.

Has anyone else proved Hotel pay tv is viable?

Yes, Computer Cinema, a division of Computer Television, has also developed a favorable market test at the Gateway Downtowner Motor Inn, located in Newark, New Jersey. It is primarily a businessman's hotel and is served by all New York television stations, as well as a large number of motion picture theatres a few blocks away. Of the hotel's 259 rooms, 120 were equipped to receive Computer Cinema.

What were the results of this test?

Sixty-five percent of all occupied rooms viewed some television during the study. Of these, 37 percent viewed and paid for a Computer Cinema film. This means 25 percent of all occupied rooms viewed and paid for motion pictures. These percentages may be understated, since rooms vacated late in the day were considered occupied; actually, the rooms were unoccupied during the viewing hours.

Hotel occupancy increased during the test period from 48 percent to 63 percent, and the hotel requested Computer Cinema to continue after the test. The hotel paid about $23 per room per month for this service.

A survey of guests indicated an overwhelming attitude favoring the system. Prior to being questioned, 60 percent of Computer Cinema viewers had not gone to a theatre for six months. While 44 percent had already seen a film, they viewed it again because of the comfort and availability of motion pictures in hotel rooms.

Computer Cinema fared well against broadcast television under all conditions. Though football did detract from the audience, beauty pageants did not. Finally, during the week that networks display their new season's programs, Computer Cinema was viewed by 25 percent of the hotel audience (compared to only 22 percent for CBS, 19 percent for ABC, 19 percent for NBC, and 15 percent for all others combined).

What is the FCC's position on allowing hotels to be interconnected via telephone company coaxial lines, cable television, and microwave?

Columbia Pictures Industries' Trans-World Communications division interconnects their hotels for pay tv using all three transmission mediums. The FCC supports these arrangements.

Is there any opposition to the hotel operators using telephone lines furnished by the telephone company?

Yes, cable operators are concerned that they could lose the hotels as potential markets for their cable services. However, the FCC has ruled in favor of Trans-World Communications in this challenge.

Have any private microwave frequencies been assigned for hotel service?

Yes. The FCC granted four applications by Columbia Pictures Industries to use private microwave frequencies (in the 12200- to 12700-MHz band of the Business Radio Service) to transmit motion pictures to hotels in Boston, Las Vegas, Dallas, and New Orleans. Columbia has applied to the FCC for use of this transmission method in cities where hotels are widely separated, where transmission by telephone lines is less desirable, and where existing cable systems are not available.

Is it anticipated that the FCC will always make all three mediums available for the pay tv service?

In 1973, the FCC ordered an inquiry into the entire subject of wire and radio transmission of motion pictures to hotels. The commission said it would consider requests by cable operators wanting to compete

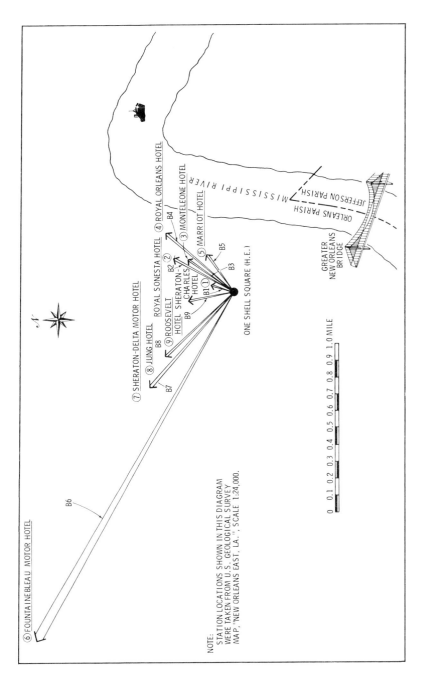

Fig. 5-1. Layout of a microwave intracity hotel system.

with transmission concerns for motion picture distribution rights to hotels. The Trans-World Communications position is that:

> It was only natural for the Commission to want to explore the various implications of a new and important communications medium.
> The principal thrust of the inquiry is not to impose limitations on Trans-World's present method of operation, but rather to explore the extent to which cable television systems may compete in providing services such as those furnished by Trans-World.

Where is the first microwave intracity hotel system being installed?

The first such system to be installed will be in New Orleans. The system will enable nine hotels to operate from a single studio, using microwave distribution (Fig. 5-1). There will be two pay tv channels and a special TEL-AD service. This special service will give information about public events in the city and offer descriptions of the current pay tv programs.

COMPUTER TELEVISION, INCORPORATED

What is Computer Television's special area of activity?

Computer Television is run by former National Broadcasting Company executive Paul Klein and is dedicated to providing random access viewing of special programing to hotels on a pay-per-view basis. Its division, Computer Cinema, was the first company to provide motion pictures on a pay-per-view basis to hotel and motel television sets. They were operational in June 1971 in Newark, New Jersey.

What is the nature of Computer Cinema's Services?

Computer Cinema has three systems: Maxess, Memory-Matic, and Instant Routing.

Maxess (Fig. 5-2) is a full two-way interactive computer-controlled system which connects to the master antenna system of the hotel. There is a room terminal installed in each guest room which provides two channels of pay tv programing, Cinema I and Cinema II. The system interfaces with a computer which bills the customer after a free ten-minute preview of the program. This system is operating in New York City and in Chicago.

Memory-Matic (Fig. 5-3) is a converted system developed by Computer Cinema. The hotel system must be hard wired or remotely read for billing purposes. This system contains no provisions for remotely inhibiting the programing to a particular room, as is contained in the Maxess system. The program can, however, be inhibited from within the room.

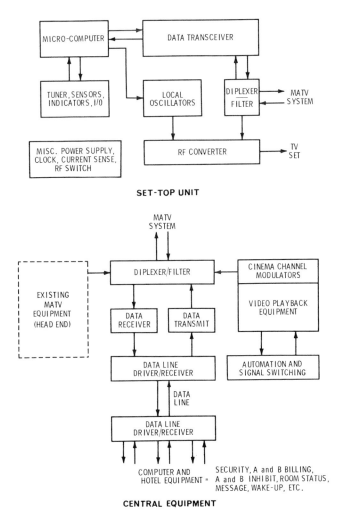

SET-TOP UNIT

CENTRAL EQUIPMENT

Fig. 5-2. Maxess system.

Instant Routing (Fig. 5-4) is the most sophisticated of the three systems. It requires hard wiring of the hotel system and provides any number of channels to the room viewer. In addition to the features of Maxess, Instant Routing provides two-way video. It can broadcast

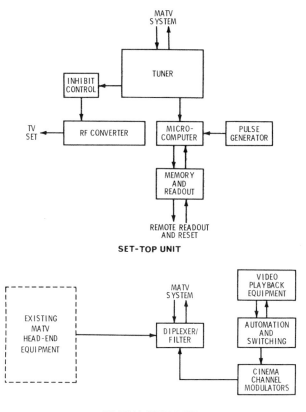

SET-TOP UNIT

CENTRAL EQUIPMENT

Fig. 5-3. Memory-Matic system.

from the room to other parts of the hotel, providing video baby-sitting and other services. In 1973 this system is operating in: Newark, New Jersey; JFK International Airport, New York, New York; Anaheim, California; Orlando, Florida; Phoenix, Arizona; and Virginia Beach, Virginia.

Are additional unused channels required on the room television set to receive the Computer Cinema service?

The Maxess and Memory-Matic systems convert two midband channels via the room converter. This is generally accomplished by using channels 3 and 4, depending on the particular area. Instant Routing provides the same capability with additional closed-circuit, two-way capability.

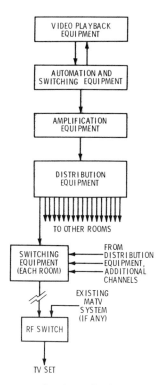

Fig. 5-4. Instant Routing system.

How are Computer Cinema's programs transmitted to the hotel guest rooms?

Using Sony video tape recorders, modified for automation, the machines are placed within the hotel or at the head end of a cable system. The signals are distributed over existing master antenna systems, and the entire system provides two-way capability.

How does Computer Cinema's concept in marketing vary from other pay tv systems?

Ten minutes of free viewing are offered so that if a hotel guest does not like the programing, he is not billed. The viewer may watch the program again the same day at no additional charge. Because of the centralized metering system (Fig. 5-5), Maxess and Instant Routing may add such features as wake-up lights, maid status, fire and theft alarms, etc. Additionally, on all three systems, Computer Cinema offers a convention package. They will do inside production and broadcast promotional material, general convention information, and items of specific interest to convention attendees.

Fig. 5-5. The computer Cinema channel selector gives the hotel guest a choice of seeing regular television, or pay tv on either of two channels, Cinema I or Cinema II. This unit has a message light and an alarm circuit.

How does the viewer know what programs are being offered on the pay tv channels?

Tent cards are on display in the rooms describing the programing on Cinema I and Cinema II; each set also has a continuously broadcasting promotional channel. A publication similar to "Playbill" is planned for the near future.

How are the pay tv channels controlled?

The systems are two way, interfaced with a computer. After a free 10-minute preview time has elapsed, the room terminal unit activates an indicator light, and a charge is added to the room bill via automatic printout (Maxess, Instant Routing) or manually (Memory-Matic). Viewing may be restricted on any channel via inhibiting switches at the front desk or in the room.

How does the viewer request the pay tv program of his choice?

The viewer tunes the tv set channel selector to a specified vhf channel (usually 3 or 4) and the Computer Cinema dial to either Cinema I or Cinema II (Fig. 5-6). If he stays tuned to one channel for more than ten minutes, an admission charge of $2.50 for that channel will automatically be added to his room bill. Once he has been billed for a motion picture, he can watch it as often as he likes until the next day, when two different films will be available.

Fig. 5-6. Computer Cinema centralized metering system for hard wiring.

What motion pictures have become available to Computer Cinema?

Computer Television has contracts with film distributors. They have relatively free access to a variety of films due to the closed-circuit nature of their hotel systems.

Will Computer Cinema book X-rated films?

Computer Cinema will show "soft" X-rated films, but will advise room occupants that they may prevent children from ordering a film by calling the front desk to inhibit either motion picture. A second call to the desk by the adult occupant of the room will restore control for viewing.

How are Computer Cinema systems sold to hotel operators?

Direct purchase includes a service charge for programing and maintenance, with a percentage of gross going to Computer Cinema. Lease plans have cost variations depending upon the number of units in the hotel or motel. The hotel owner receives 25 to 40 percent of gross, depending upon how many tickets are sold per night. Computer Cinema will also consider a joint venture with a hotel, in which case profits are split.

What is the price schedule for Computer Cinema's Instant Routing System?

See Chart 5-1.

PHILCO-FORD / THEATREVISION

What is Philco-Ford Corporation doing in pay tv?

Philco-Ford Corporation, through its Tele-Sound Division, has entered into a cooperative project with TheatreVisioN to bring the TheatreVisioN pay tv system into thousands of hotel rooms across the United States.

How will Philco-Ford/TheatreVisioN operate?

A pilot hotel system, the components of which are being assembled by the Philco Corporation, has been installed in the Sheraton Penn-Pike Motor Inn, Ft. Washington, Pennsylvania. This system utilizes the encoding and decoding devices of the TheatreVisioN cable pay tv system.

One hundred thirty-five rooms were equipped with the decoding units. Hotel guests may watch two motion pictures nightly, with films alternating on odd and even days of the month.

How will Philco-Ford deliver these programs to the hotel rooms?

Tele-Sound (headed by Mr. Barney Gardner), in cooperation with the engineering division of Laser Link Corporation, has manufactured a two-channel program console (Fig. 5-7). This console is completely automated and will transmit through a conventional master antenna system in the hotel. The system will utilize four Sony U-Matic video-

Chart 5-1. Price Schedule for Computer Cinema Instant Routing System

	1-76 Rooms	77-128 Rooms	129-256 Rooms	257 Rooms and over
Room installation Fee	$150 per room	$24 per room*	$24 per room*	$24 per room*
Video Origination Equipment—5 Sony U-matic Players (automated)	$6000	$3000	None	None
Head End Switching Control Unit and Billing Panel	$4000	$2000	None	None
60-Month Term. Fee on a per-Room, per-Month Basis	None	$4	$4 for rooms 1-128 $3 for rooms 129-256	$4 for rooms 1-128 $3 for rooms 129-256 $2 for rooms 257 and over
Cost per Film Title	$100	$50	None	None
Hotel/Motel Revenue	25% of gross from first 15 tickets per day per 100 rooms; 40% of gross from all tickets sold in excess of 15 tickets per day per 100 rooms (computed on an average daily basis per calendar month).			
	(Assumes 76 rooms)	(Assumes 128 rooms)	(Assumes 256 rooms)	(Assumes 400 rooms)
Hotel/Motel Projected Annual Revenue at occupancy of 70%	$3,295	$5,556	$11,112	$17,338
80%	3,988	6,716	13,432	21,988
90%	4,683	7,884	15,768	24,638

(Assume 25% of occupied rooms buy one film per night at $2.50 per movie.)

Computer Cinema is responsible for all maintenance.

* Deposit applied against last month of term of agreement.

cassette players which will serve as the origination equipment for programing. TheatreVisioN will buy, book, and deliver current films and electronic tickets specifically coded for each of the films to be offered.

Fig. 5-7. Barney Gardner, manager of the Tele-Sound Division of the Philco-Ford Corporation (left) and Stanley Spiegelman, project manager for TheatreVisioN, reviewing the two-channel console which transmits two coded pay tv programs through the hotel master antenna system.

How will the programs be sold to hotel guests?

Posters in the lobby and elevators and tent cards in the rooms will indicate the pictures to be shown during the month. Guests may obtain tickets from a dispenser at the registration desk. Tickets may also be delivered to the rooms and charged to the guest's bill. This simplifies problems incurred with computer interface, telephones, etc. Once the

Fig. 5-8. Will Baltin, of the executive staff of TheatreVisioN, operating a decoder.

guest inserts the ticket into the decoder (Fig. 5-8) he may view the motion picture any time for the following twelve hours.

What is planned beyond the pilot system?

It is proposed, following a 90-day test of the system at the Penn-Pike, to expand into hotel rooms presently being serviced with television receivers by Philco. In 1973, over 3000 hotels and motels coast to coast have Philco receivers in their guest rooms. It is proposed to provide TheatreVisioN in at least 50 to 60 percent of the presently equipped hotels.

What type of programing will be offered by Philco in hotels?

Films and sporting events in current circulation in theatres and sports arenas will be carried on the hotel systems. A guest will pay a price of $3 for each program for a viewing period of up to 12 hours.

What advantages are offered to the hotel/motel operator?

Philco-Ford finances the entire installation of the TheatreVisioN system and the hotel operator, without making an investment, will receive a percentage of the ticket price. Philco's 3000 national service agents are available to keep the system working and to distribute new cartridges, tickets, and promotional materials. The Philco servicemen will also remove the destroyed tickets that have accumulated in the TheatreVisioN decoder boxes.

PLAYER'S CINEMA SYSTEMS, INCORPORATED

What is Player's Cinema Systems' special area of activity?

Player's Cinema Systems, founded in mid-1970 by Jerry Grote and Bud Harrelson, is a division of Player's Group Companies. Player's Cinema is a leading distributor of hotel/motel in-room film systems.

What is the nature of Player's Cinema Systems' services?

There are two basic systems offered on a per-program charge basis to the individual room viewer. The hotel/motel operator may pay for the service on a per-room basis, in which case Player's supplies equipment and software; or he may buy the equipment and rent software from Player's.

The Player's "Movie Box" (Fig. 5-9) is a self-sufficient system that works independently of television. This system, exclusive with Player's, enables guests to choose from a wide variety of films. A guest can begin watching the film any time by just pushing the START button. Should he be interrupted by the telephone, etc., he can stop the film at once and restart it later at the same point without missing any of it. Since the Movie Box is completely self-contained, with a rear projection unit, nothing can interfere with the picture.

The Player's "Video Box" system consists of a console that can be placed in a convenient location. This unit is connected to the base of the master antenna system for transmission to all existing tv sets within the hotel/motel. The player is loaded once, and the films automatically go on at set intervals during the day. After insertion of film cassettes, no operator attention is required. The system is completely a hands-off operation. The Video Box system can be used to offer free films to guests, or as a revenue-producing system. If the operator prefers to charge for the films, the basic system can be supplemented with the capability of customer billing. This system is also installed without changing the hotel's present wiring. Guests can watch a free preview of the film to decide if they want to see it. After a preset time,

Fig. 5-9. New York Mets' Bud Harrelson, vice president (left), and Jerry Grote, vice president of Player's Cinema Systems, with the Cassette Mini-Theatre ("Movie Box").

a centrally located console automatically produces a hard copy indicating which room is to be billed.

Are additional unused channels required on the room television set to receive the Player's Cinema Systems' service?

No, the television system is closed circuit. The Video Box can utilize any unused vhf channel.

How are programs transmitted to the hotel's guest rooms?

The Movie Box is a totally self-contained unit requiring no transmission facilities. The Video Box is connected directly to the hotel's master antenna system.

How does the Player's Cinema System concept in marketing vary from other pay tv systems?

The Video Box system does not demand per-program payment from the hotel guest. It is not "pay tv" insofar as the guests are concerned.

The philosophy of this company is that, if the guests watch the programs, in-room motion pictures will become as commonplace in hotels as air conditioners or tv sets. In addition to feature films, Player's Video Systems can be used for convention use (training films, etc.), private showings, and sporting events.

How does the viewer know what programs are being offered on the pay tv channel?

"Film menus" are distributed to the guest rooms. This literature lists and describes the programs and gives the rating of each film. Promotional items are also displayed in the lobby of the hotel.

How does the viewer request a pay tv program?

The programs can be made available, without charge, on the Video Box and run continuously on the tv screen. This system is offered as a feature of the hotel in its list of services. The customer must contact the desk to request a film on the revenue-producing Movie Box system.

How are the pay tv channels controlled?

This varies with the system and the individual hotel operator's philosophy. On a per-program basis, the guest is charged at the hotel's billing center. If films are offered as part of the hotel's service, there is no charge.

What films have become available to Player's Cinema Systems?

Primarily G- and GP-rated motion pictures from major studios.

Will Player's Cinema Systems book X-rated films?

Not on the Video Box. X-rated films will be shown on the Movie Box, since more controlled viewing can be exercised on that unit.

How are Player's Cinema Systems sold to hotel operators?

Possible options offered to hotel operators are: rental, lease, or outright purchase.

Are there any other pay tv proponents for hotels and motels which use playback equipment alone?

Yes. Cartridge TV of New York City plans to offer their cartridge rental network to hotels. In 1973 they had over 200 film titles and plan to rent their cartridges and cartridge-playback equipment for $2 to $6 a showing.

TELEBEAM CORPORATION

What is Telebeam Corporation?

Telebeam is a privately held corporation that was founded in 1970. The immediate goal of the company is to develop as many as ten programs simultaneously. This is a two-way communication package including: room security, room status, point of sale terminals, and information and reservation services for hotel use. A pay tv and security system for apartment or CATV users is also available.

Where is Telebeam marketing its system?

The initial installation is in the Americana Hotel, New York City. Contracts have been signed, or are in various stages of negotiation, with other key hotels, motels, and apartment houses throughout the United States and abroad. The company projects well over 25 thousand subscribers in New York City by the end of 1973 and well over 200 thousand nationally by 1975.

What is the nature of Telebeam's services?

The system is designed primarily for the transmission of pay tv programs directly from a central station to hotels and apartment houses in urban areas. For subscribers outside of a direct transmission area, Telebeam has developed and is marketing a complete on-premises system. Utilizing sophisticated electronic and computer equipment, individual viewers can select a program from a number of programs being transmitted simultaneously.

How does Telebeam's pay tv concept vary from other proposed pay tv systems?

The Paytel entertainment service is a totally different approach than the systems being used by other proponents. Depending on the locale, the hotel or apartment will be furnished with its own head-end transmission center (Fig. 5-10). Or Telebeam will provide a link between a central transmitting station and the receiving location. In either instance the existing MATV system in the building is used to transmit the programs to the specific rooms. At any time, each room occupant can choose from as many as ten different programs. These can be current films, operas, ballets, or live sporting events. The Telebeam computer automatically routes a selected program to each viewer. At the same time, the computer records the transaction for billing the viewer. In hotel installations, a high-speed teletype provides a hard copy of the information, and the cost is added to the guest's bill.

Fig. 5-10. Telebeam Corporation's Paytel system head-end transmission center and monitoring panel.

How does the viewer know what programs are being offered on the pay tv channels?

The viewer is acquainted with Telebeam by a printed guide listing programs being shown, times of performances, and cost. Further, there is a preview channel which previews the shows being exhibited, gives performance times and prices, and provides instruction on usage of the system.

How does the Telebeam viewer request a pay tv program of his choice?

Using the Telebeam Command Control Unit, the viewer sets the dial to the pay tv channel of the program of his choice. He then presses a BUY button and a PUSH TO ORDER button on the control unit. The selected program will then appear on the tv set. The viewer can order and view a program at any time he wishes. The billing process takes place ten minutes after viewing begins. The viewer may cancel selection by simply pressing the CANCEL button before ten minutes have elapsed.

What are the other services which are made available via this new transmission means?

As part of Telebeam's total communications package, the company provides a rigid room-access security system (Gardtel), a room status system (Roomstat), Point of Sale Terminals, and an information and reservation service (Datatel). (Fig. 5-11.)

The Gardtel security system is able to monitor 2000 doors in a hotel or apartment house once a second. Each hotel guest or apartment occupant is given a coded card. The code on each card is fed into the Telebeam computer, which then knows that a specific card corresponds to a specific door. Upon entering, using the normal keylock, the occupant places his card into the card reader situated on the door or on the control unit. If the room is entered without insertion of the proper card, an alert is flashed over the Gardtel security monitor. The exact location of an unauthorized entry appears on the

Fig. 5-11. The Paytel system head-end transmission center and monitoring panel with the Datatel, Gardtel, and Roomstat equipment. The room terminal unit is shown atop the television set.

monitor, plus personal identification. A security employee immediately telephones that room. If there is no answer, someone is dispatched to the room. If the telephone is answered, and if the proper person is in the room, he is reminded that he must insert his card into the card reader. If the proper card is not inserted, then appropriate action is taken.

When a guest checks out of the hotel, the coded card corresponding to that room is erased from the computer's memory. The next guest assigned to that room receives a differently coded random card. Maids are issued a different worker card each day for their specific rooms. In the home, the Gardtel system may be utilized for front and rear door entry or access to a garage. The Gardtel system has been deemed so effective that a major insurance underwriter allows a 10-percent reduction in their theft policy for those apartment dwellers that have this system installed. Fire and smoke sensors are optional with the Gardtel unit.

Utilizing the most advanced computer technology, the Telebeam Roomstat system provides up-to-the-minute room status information. The Roomstat video monitor displays the number of rooms sold, reserved, available, and available but not ready for occupancy. The Roomstat system includes a computerized "wake-up" and message display for each room.

Included in the Roomstat system are the Automatic Charge Terminals. These terminals were designed to provide an effective method of completely eliminating the frequent "charge after departure" problems now plaguing many hotels. This unique feature enables cashiers to determine whether a person charging a purchase to his room is a registered guest. Additionally, charges entered at the purchase terminals are immediately transmitted to the hotel's main billing facility, assuring that all charges made by a guest will appear on his final bill.

The Datatel information and reservation service is available where and when demand exists. This is a free service which allows the viewer to communicate via his television set and obtain detailed information on: restaurants (location, telephone numbers, menus, etc.); airline and other transportation schedules; points of interest; shopping catalogues; etc. After obtaining the desired information, the viewer using the Telebeam command unit is able to reserve tickets or make purchases from stores. He receives confirmation of his purchases over the tv set.

How many rooms must a hotel have to get Paytel?

Telebeam usually installs Paytel in hotels that have 150 rooms or more.

What does a Paytel installation cost?

Telebeam installs and maintains the Paytel system at its expense. It uses the hotel's existing MATV system for distribution of Paytel services.

CONCLUSION

Who else is currently designing two-way pay tv systems for hotels and motels?

Primary Entertainment Corporation (a joint venture between Bell and Howell, Primary Medical Communications, and 20th Century Fox), Motorola, and TransCom Productions are currently designing these systems.

What are Primary Entertainment Corporation's methods?

In the PEC system, the viewer turns on the tv set to an unused channel(s) and watches ten minutes of a motion picture free. If he watches after ten minutes, he is automatically billed on a hard-copy printout. The system currently anticipates the use of two channels, each showing a different film, with programs alternating every other day. They anticipate 24 to 28 films will be offered each year.

Where is PEC being tested?

In 1973 systems should be operational in a hotel near Brentwood, California. They also plan to make some hospital installations.

What are Motorola's systems methods?

Motorola has reported plans for a wireless two-way communication system, employing the hotel's existing MATV systems. Like Philco, RCA, Wells TV, and others, Motorola leases tv sets to hotels and motels. When the subscriber operates the pay tv in the Motorola system, billing information is provided via a tv set. The room numbers of those watching programs are screened over the tv set at the billing location.

Where has the Motorola system been tested?

Motorola has completed a test of its system at Des Plaines, Illinois.

What are TransCom Productions' system methods?

In the TCP system, each hotel/motel room contains a selector unit, which is activated by the insertion of a key into either channel A or B. TransCom uses super eight projectors with color tv cameras to deliver the programs on the MATV system. Billing information is fed through existing wiring to a central location. The guest is then billed at checkout time. They plan to show films three times a day, at an average cost of $3 per program. Each film can be viewed continuously for that price.

Where is TransCom Productions being tested?

TCP plans tests in hospitals, motels, and hotels in California during 1973.

How do hospital systems differ from hotel systems as a market for pay tv?

From the pay tv viewpoint, there is very little difference. In hospitals, all television sets are on a master antenna system; the same arrangement is used in hotel systems. The major proponents in this field are: Wells National Services Corporation, New York City; Sylvania Hospital Services Division, Waltham, Massachusetts; and Electronic Associates, a division of RKO General, New York City. About six other systems are strong on a regional basis.

Is pay tv adaptable to hospital systems?

Yes, but only minor inroads have been made in the hospital market by 1973. The main resistance has been from hospital managements; they are concerned about the affect any new service might have on their daily schedule. Also, it is difficult to install even a small studio on some hospital premises. Additionally, the hospitals do not want to participate in any billing procedures.

It is anticipated that the hospital market for pay tv will be slow in growth. The service will have to be accomplished in such a manner that it relieves the patient's boredom without burdening the hospital with additional administration and monitoring assignments.

Hotels, being competitive businesses, are anxious to gain customers. Hotels are only 65-percent occupied; therefore, they desire incentive services.

Chapter 6

Cable Pay TV Systems

INTRODUCTION

What are the differences between hotel/motel pay tv and cable pay tv systems?

Most cable pay tv systems are not designed specifically for the hotel and motel requirements. When the hotel/motel systems (including the master antenna system and the tv sets) are under the control of the pay tv proponents, a lower level of security can be used. The average hotel guest only spends a few days in the room, and the pay tv program is a one-time charge. The home instrument is sitting on the subscriber's premises and there is a charge for every program.

Pay tv over cable systems will have to provide reasonable video and audio security so that misuse will not take place in the future. While uncoded pay tv programing in midband (108 to 174 MHz) channels A through I may be satisfactory in 1973, most pay tv proponents do not believe it will be satisfactory in 1975. Some new television sets are being equipped with midband channels A through I on the set tuner; this would allow pickup of cable pay tv programing, broadcast in the midband, without paying for the service. Also, low-cost midband converters and kits could become an item with catalog houses by 1975.

What degree of scrambling is needed for cable pay tv?

Cable pay tv encoding does not have to be as complex as over-the-air techniques, but it must be sufficiently secure to require sophisticated technical knowledge to overcome the encoding/decoding format. Obvious misuse of this type of system is subject to legal action and damage claims by the cable pay tv proponents, most of which have patents and/or licensing agreements. Cable pay tv encoding/decoding system designers do not try to achieve high-level security. It is too costly, and they are not concerned with being defeated by the super technician. The technician may be subject to costly prosecution if he illegally alters the system and offers his service to others.

Do any proponents design their systems for both cable pay tv and hotel/motel pay tv markets?

Yes. Athena Communications, TheatreVisioN, and Trans-World Communications offer systems for both the hotel and cable television markets. These systems will be described before those systems which are offered only to the cable television market.

The hotel/motel system, under the full control of Philco-Ford Corporation's Tele-Sound Division, requires less codes and no scrambling. Anyone who brings his own converter (and jumper cables) to a hotel room can see the program free. This pratice is, of course, so rare that it would not be economical to take measures against it.

Which systems were offered in 1973 exclusively for cable pay tv?

The following companies offered pay systems exclusively on cable television in 1973; Cinca Communications; Gridtronics, Inc; Home Box Office, Inc; Home Theater Network, Inc; Jerrold Electronics Corporation; Optical Systems Corporation; Skiatron Electronics & Television Corp; TelePrompTer/Magnavox Premium Television.

However, it should be noted that nearly every two-way system designed with centralized metering has pay tv potential. These systems will also be described in this chapter.

Who are the known proponents of cable pay tv systems, including hotel systems?

Athena Communications Corp.
One Gulf & Western Plaza
New York, New York 10023

James Ragan, President

Cinca Communications
9229 Sunset Boulevard
Los Angeles, California

Robert Breckner, President

Computer Television, Inc.
Fifteen Columbus Circle
New York, New York 10023

Paul Klein, President

Gridtronics, Inc.
630 Fifth Avenue
New York, New York 10020

Alfred R. Stern, President

Home Box Office, Inc.
Time/Life Building
Rockefeller Center
New York, New York 10020

Charles F. Dolan, President

Home Theater Network, Inc.
1880 Century Park East
Los Angeles, California 90067

Richard Lubic, President

Optical Systems Corp.
11255 Olympic Boulevard
Los Angeles, California 90064

Geoffrey M. Nathanson, President

Player's Cinema Systems, Inc.
245 Newtown Road
Plainview, New York 11803

Bud Harrelson, Vice President

Primary Entertainment Corp.
122 East Forty-second Street
New York, New York

Michael Barnett, President

Skiatron Electronics & Television Corp.
30 East Forty-second Street
New York, New York 10017

Arthur Levey, President

Telebeam Corp.
122 East Forty-second Street
New York, New York 10017

Sy Grodner, President

TelePrompTer/Magnavox
Premium Television System
50 West Forty-fourth Street
New York, New York

Don Witheridge, Dir. Public Relations

TheatreVisioN, Inc.
641 Lexington Avenue
New York, New York 10022

Dore Schary, President

Trans-World Communications
711 Fifth Avenue
New York, New York 10022

William Butters, Vice President

Are there any others?

Yes. Read "Two-Way Systems with Pay TV Potential." These systems are in an experimental phase but may be fully developed and in production by the time of publication of this text.

ATHENA COMMUNICATIONS CORPORATION

What is Athena Communications Corporation?

Athena Cablevision Corporation is the forerunner of the Athena Communications Corporation. Athena Cablevision was first organized in 1968 by Gulf & Western Industries. At that time, Athena Cablevision was formed to succeed to all of G & W's CATV operations in the United States. Several years later (December 1971) G & W organized the Athena Communications Corporation. This second operation was organized to acquire from G & W all of the capital stock of Athena Cablevision. Athena Communications Corporation then became an independent, publicly held corporation on September

TEXAS STATE TECHNICAL INSTITUTE
ROLLING PLAINS CAMPUS – LIBRARY
SWEETWATER, TEXAS 79556

8, 1972. It is essentially a holding company. Athena Cablevision is now an operating subsidiary of Athena Communications.

What is Athena's special area of activity?

Athena, through its cablevision subsidiary, has been active in community antenna television systems since 1968. Athena presently operates 16 systems with over 60 thousand subscribers. This operating area serves 33 communities in seven states. Athena also holds franchises in six additional market areas (a total of 450 thousand more tv homes). When construction is complete in these additional areas, Athena will rank as one of the ten largest CATV companies in the United States.

In the pay tv field, Athena has developed and applied for patents on a low-cost (hard-wire principle) hotel theatre system. They have also patented a low-cost scrambling system, called EnDe-Code, for pay cable operations.

What is the nature of Athena's hardware service?

In cable systems, Athena is concerned with installation for its own properties.

Athena sells and leases EnDe-Code scrambling systems to any interested pay cable operator. Development of the patented EnDe-

Fig. 6-1. Athena Corporation's Hotel Theatre System Guest Unit.

Fig. 6-2. Athena Corporation's Hotel Theatre System centralized metering panel and control unit.

Code system began on an experimental basis in 1969. At that time, the system was used in the "on-line" operation of a system in one location, and it used production hardware.

For hotel/motel pay tv, Athena has developed, and is presently marketing, the "Hotel Theatre System" (Fig. 6-1). Five of these systems are now operating, and more are under construction. All hardware in HTS systems is production hardware. A showcase system is located in the O'Hare International Tower Hotel, Chicago (Fig. 6-2).

What is Athena's programing procedure?

Athena has standard CATV services for local and imported signals. Local origination and public-access channels are offered in its own systems. Athena is developing other specialized programing for its own systems and for syndication.

In the pay cable area, Athena offers feature films and specialized programing for EnDe-Code systems that are leased from, purchased

from, or operated by the company. Athena has a contract for telecasting the St. Louis Blues home games on a subscription channel
in Jefferson City, Missouri. This is the first pay cable contract ever
made with a major league team. Athena Sports Network produced
and syndicated the NHL highlights for the 1972-1973 season. This
program was shown over a 100-station network.

Current feature films and special event programs (concerts, sports,
etc.) are offered to hotel pay tv subscribers by Athena.

Are additional unused channels required on the home television set to receive Athena's service?

The EnDe-Code system in pay cable tv utilizes existing channels
where available. If no existing channel is available, additional channels are obtainable through the use of any make converter. Also, the
EnDe-Code decoder may function as an entirely separate device.

The HTS system for hotel pay tv utilizes spare channels when
available. If no spare channel exists, HTS blocks out off-the-air
signals on the hotel MATV when the guest selects the program
channel. Depending upon the size of the hotel, the HTS system utilizes
either one or two channels. The O'Hare International Tower Hotel is
a two-channel system.

How does Athena's pay tv concept vary from other proposed pay tv systems?

The EnDe-Code system for pay cable tv provides controlled access to
any CATV channel on a subscription basis. The system makes is possible to encode any tv channel, so that standard tv sets receive
scrambled pictures and no sound. The decoder restores the picture
and sound to the tv set. EnDe-Code's advantages are: security on any
channel via the sophisticated scrambling/unscrambling systems; push-
button selection or normal or encoded mode on encoder, and low cost.

The hotel owner buys or leases the origination hardware; programing
and control equipment is provided by Athena. Revenue from the
programing is split on a percentage basis. The system in a hotel utilizes
a "remote selection" button on the decoder. The system is locked
out until program time; it is then unlocked, and fifteen to twenty
minutes of the film are shown free of charge. After this period, a
cut-in advises the viewer that the button on the decoder must be
pushed again to continue viewing the program, at a charge. If this
button is not pushed, the program is locked out of that particular set.
This avoids accidental selection. A telephone is used only if the
customer wishes to see a program in progress, after the lockout has
occurred. The operator has special controls to permit the guest to

view the film. The small (one-hundred room) hotels utilize manual billing. A lamp will light up on the console board when the pay system is being operated by a guest. In the medium-sized (three-hundred room) hotels, a printer will scan and provide a printout bill and assign it as a room charge. In the large (one-thousand room) hotels automatic interfacing with a room billing computer readout is provided.

How will the viewer know what programs are being offered on the pay tv channels?

Athena uses advertising, monthly lists of programing, and home mailers to inform pay cable viewers of its program fare.

For hotel pay tv systems, Athena provides "tents," lobby promotions, previews of films on the individual sets, and timetables in the guests rooms with short explanations of the films offered and their ratings.

How are the pay tv channels controlled?

Pay cable channels are controlled on a subscription channel basis. Programs are scrambled at the cable system head end. A subscriber with a decoder can tune in at any time. Subscribers are billed monthly in advance.

Centralized metering (manual or automatic) is the billing process used in hotel pay tv. The number of rooms in a specific hotel determine which billing method is used.

How does the Athena viewer request a pay tv program of his choice?

This procedure has been described earlier. Briefly, pay cable channels are selected remotely by a device installed on the television set in his home; in hotels, the guest pushes a button on his room installation.

Will Athena use X-rated films?

No.

When will EnDe-Code become available to the home CATV subscriber?

The first test system will begin by mid-1973 in Jefferson City, Missouri. Other negotiations are in progress, but no other locations are firm.

What does a CATV operator do to carry Athena on his system?

1. Prospective pay cable operators should contact:
   Abraham M. Reiter
   Vice President, Engineering
   Athena Communications Corp.
   P.O. Box 2127
   Norwalk, Connecticut 06854
2. Hotel/motel pay tv information may be obtained from:
   Ronald L. Haskell
   HTS Sales Manager
   Athena Communications Corp.
   1 Gulf & Western Plaza
   New York, New York 10023

What are Athena's responsibilities to the cable operator?

Athena provides programing, encoder, and decoders for pay cable operators. The company will install only the encoder. Promotion, advertising, and marketing is done by the Athena group.

Does the cable operator have to buy the EnDe-Code converters or can he lease them?

Larger systems may be permitted to lease the converters.

Do Athena's converters have any other special features?

1. The converter is a one-way unit in pay cable systems.
2. The converter is a two-way unit with theft and room status options for hotel pay tv operations.

CINCA COMMUNICATIONS

What is Cinca Communications?

Cinca Communications is an organization run by Dimitri Villard and Robert Breckner, former Times-Mirror cable executive.

What is Cinca Communications' special area of activity?

Cinca plans to furnish pay tv films over leased television channels under the trade name of Cinca Channel One. Cinca's billing will be accomplished via credit card or direct mail.

Are additional unused channels required on the home television set to receive Cinca's services?

In all instances, Cinca plans to use Oak Industries converters. They will be modified with a key feature for pay tv purposes, so that the adult in the home will control the programing.

How does Cinca's pay tv concept vary from other purposed pay tv systems?

Cinca will be promoting a $7.95 installation charge and thirteen weeks of films for $19.50. The films will be run, as either single or double features, three times a day starting at 3:00 P.M. The credit card billing is also considered unique to this system.

How does the viewer know what programs are being offered?

Cinca has a video program which it promotes into the cable system. Advertising will also be done in the local newspaper.

How are the pay tv channels controlled?

The channels are controlled by use of the home converter and the key feature contained in the unit.

How does the Cinca viewer request a specific pay tv program?

The viewer pays a flat rate for the Cinca service; therefore, all programing is available to him.

What programs are booked for Cinca Communications?

Cinca has arranged for film products from several major studios.

When will Cinca Communications' service become available to home CATV subscribers?

Cinca started its service in March 1973 at Long Beach, California. They plan to have systems in Escondito and San Clemente operational before the summer of 1973.

Will Cinca use X-rated films?

No, but the company will show R-rated films with the key feature under parental control.

What are Cinca Communications' responsibilities to the cable operator?

Cinca leases a channel and pays a fee for this lease. They furnish programing, hardware, and software to the cable operator.

GRIDTRONICS, INCORPORATED

What is Gridtronics, Incorporated?

Gridtronics is a subsidiary of Warner Cable Systems which is part of Warner Communications. Warner Cable is one of the largest multiple systems operators in the cable television industry. Gridtronics was introduced to the industry by Alfred R. Stern, a leading executive in the cable field.

What is Gridtronics' special area of activity?

Gridtronics offers a special converter for the individual subscriber. They have an engineering package, for the cable tv operator, which provides all the technology needed to operate a television studio feeding into a cable system.

What is the nature of Gridtronics' service?

Gridtronics, as a subsidiary to a cable system organization, offers services not only to its own organization, but also to the cable industry in general.

Are additional unused channels required on the home television set to receive Gridtronics' service?

Yes. Four channels are being offered on the Gridtronics system.

How does Gridtronics' pay tv concept vary from other proposed pay tv systems?

Gridtronics utilizes the midband and inverts the position of the aural and visual carriers to provide effective scrambling with its encoding equipment.

Conversion is accomplished by a crystal-controlled block converter. This converter beats the selected program into the vhf channel chosen for television set use in the specific area. One, two, or three channels are receivable in isolation or in combination. This is done by setting the crystal oscillator frequency.

The Gridtronics units have a key control which leaves the subscriber terminal under the control of the parents. They may restrict the viewing of R-rated films or other programs which will be offered.

How does the viewer know what programs are being offered on the pay tv channels?

The subscriber will be informed of what programs are to be shown on the Gridtronics system via program guides and the marketing efforts of the cable operator.

How are the pay tv channels controlled?

The channel viewing is controlled by the home subscriber with his key.

How does the Gridtronics viewer request a pay tv program of his choice?

Programs, over this system, are offered on a flat-rate basis. In 1973, Gridtronics rates were between $5.00 and $7.50 per month. The subscriber will receive 104 films a year: two per week, shown on alternate days, four times a day. In addition, bonus programs (travelogues and cartoons) will be offered.

Will Gridtronics use X-rated films?

They will not use X-rated films.

When will Gridtronics become available to the home CATV subscriber?

The first ten systems are scheduled to be placed in installations owned by Warner Cable Systems by mid-1973.

What does a CATV operator do to carry Gridtronics on his system?

Gridtronics offers a total pay tv service to the cable operator on a turn-key basis.

Does the cable operator have to buy the converters or can he lease them?

The cable operator, in early 1973, was offered the opportunity to buy the Gridtronics product. Purchase was recommended because the operator could soon absorb equipment cost through his increased revenue potential.

Does Gridtronics' converter have any other special features?

Gridtronics is committed to the keylock feature. Even in their earlier systems, where Jerrold converters were used without scrambling, a key control was necessary to make the converter operational. The converter is also proposed for use with a rate card (Fig.6-3), depending on the services which may be offered by the cable industry.

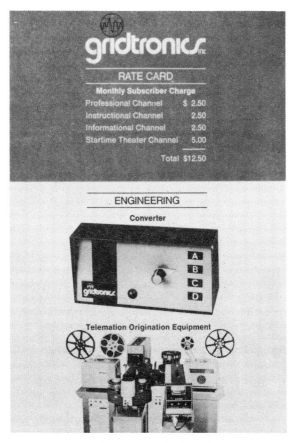

Fig. 6-3. Gridtronics' rate card.

HOME BOX OFFICE, INC.

What is Home Box Office, Incorporated?

Home Box Office, a subsidiary of Sterling Communications, was formed in 1972 to provide pay programing for cable television.

What is Home Box Office's special area of activity?

At the moment, Home Box Office is involved primarily in providing films and live sports via microwave or AT&T long lines. Additional general programing is offered from the Time-Life Library of films. Madison Square Garden has signed a five-year contract with Time to carry all pay tv telecasts.

Are additional unused channels required on the home television set to receive the services?

In most instances, midband channels will be used. The service is currently operating in Allentown and Wilkes-Barre, Pennsylvania, where converter channel "H" is used to receive the programs.

How does Home Box Office's pay tv concept vary from other proposed pay tv systems?

Home Box Office does not want to become involved in systems utilizing cards, tickets, etc. It feels that this will become obsolete shortly. The company is using one central location in New York City for broadcast purposes. The cable operator pays for transmission from the broadcast point to his head end.

How does the viewer know what programs are being offered?

A monthly program schedule (Fig. 6-4) is sent out to subscribers. In Wilkes-Barre, the schedule is carried in the local newspapers. Program reminders are carried on the local origination channels of participating cable systems.

NEW YORK KNICKS

Saturday, February 3	8:00	vs	Cleveland Cavaliers
Tuesday, February 6	7:30	vs	Los Angeles Lakers
Saturday, February 10	8:00	vs	Detroit Pistons
Wednesday, February 14	7:30	vs	Chicago Bulls
Saturday, February 17	8:00	vs	Philadelphia 76ers
Tuesday, February 20	7:30	vs	Portland Trail Blazers
Saturday, February 24	8:00	vs	Buffalo Braves
Tuesday, February 27	7:30	vs	Boston Celtics

NEW YORK RANGERS

Sunday, February 4	7:00	vs	Atlanta Flames
Wednesday, February 7	7:30	vs	N.Y. Islanders
Sunday, February 18	7:00	vs	N.Y. Islanders
Sunday, February 25	7:00	vs	Minnesota North Stars
Wednesday, February 28	7:30	vs	Chicago Black Hawks

Immediately following these games, we'll present a 10-minute demonstration of the new TIME-LIFE VIDEO SPEED READING SYSTEM featuring Dick Cavett as the instructor.

The entire course of eight 40-minute lessons on video cassettes can be taken in your own private study booth at the TIME-LIFE VIDEO CENTER, 6th Avenue and 50th. You can significantly increase your reading speed and improve your comprehension for just $90 — less than the amount you'd probably pay for a conventional speed reading course. And pay for everything on your American Express, Master Charge or BankAmericard.

To schedule your first lesson, just call 556-3210 or come to the Center.

And while you're there, you can preview the wide selection of other video cassette programming featuring Jack Nicklaus, Julia Child, Leonard Bernstein—and ranging from business to entertainment to health to education to sports and leisure—that TIME-LIFE VIDEO offers right now.

We hope we'll be seeing you—and you'll be seeing us—soon.

Fig. 6-4. How Home Box Office combines entertainment and educational pay tv into its cable subscription service.

How are the pay tv channels controlled?

By use of the converter. There are no elaborate security features.

How does the Home Box Office viewer request a pay tv program of his choice?

There is no choice since Home Box Office is a monthly subscription pay tv system. Thirty percent of the homes in Wilkes-Barre have signed up for the sports and motion picture programs. Sample tests reported 75 percent would prefer to watch at home; less than 10 percent would rather go to the theatre.

What programs are booked for Home Box Office?

Sporting events (Fig. 6-5), feature films, and general programs are offered from the corporation's access to an excellent film library.

Fig. 6-5. Home Box Office is providing a new career opportunity for former heavyweight champion Floyd Patterson (left). He is a featured commentator for boxing telecasts from Madison Square Garden. Veteran Don Dunphy handles the "blow-by-blow."

When will Home Box Office become available to the home CATV subscriber?

The company has been operating since 1972 in Wilkes-Barre, Pennsylvania. It plans to go into Allentown starting February 1, 1973, and progress in other areas will soon follow.

Will Home Box Office use X-rated films?

No. Due to little control over the channel, X-rated films will not be offered at this time. The company plans to continue using some films with an R-rating.

What does a CATV operator do to carry Home Box Office on his system?

The cable operator must pay for transmission of programing from the company's broadcasting center in New York City to his head-end location. He must provide all terminal equipment and converters obtained from a source other than Home Box Office.

What are Home Box Office's responsibilities to the cable operator?

Home Box Office provides early selling, promotional material, continuous monthly schedules of programing, and cooperative advertising to the cable operator.

Does the cable operator have to buy the converters or can he lease them?

The cable operator may buy or lease the converters at his option from an outside supplier. Home Box Office is strictly in the software business.

HOME THEATER NETWORK

What is Home Theater Network, Incorporated?

Home Theatre Network is a Los Angeles-based company conceived by cable tv veteran Richard Lubic, backed by J. Ronald Getty, and Calvin Johnston, chairman of Property Research. Johnston and Lubic own 75 percent; 25 percent is held by the Getty interests.

What is Home Theater Network's special area of activity?

Home Theater Network, formed in 1972, plans to offer its pay tv services to cable tv operators in selected markets in 1973 and to expand nationally shortly thereafter.

What is the nature of Home Theater Network's service?

Basically, the Home Theater Network system works as follows:

WHAT'S NEW FOR THE HTN VIEWER

The Home Theater Network will add 2 new channels to the subscriber's present tv set. These two living room theaters (the "Hollywood" and the "Premiere") will offer a wide variety of top flight entertainment nightly.

To supply the film demands of these new theaters, HTN will use selected product from both Hollywood and European studios . . . a minimum total of 125 new or current motion pictures a year. Supplementing these movies will be Broadway plays, concerts, rock festivals, sports, and many other special attractions. Viewing times and dates for individual programs will be rotated, insuring the viewer of seeing his selection when he wants it.

HTN KEEPS SUBSCRIBERS INFORMED

In the entertainment industry, "pre-selling" the audience is most important. Thus HTN's subscribers will receive periodic program guides, depicting play dates, times, and associated prices of the events available in their living room theaters. Additionally, a daily, early evening show called "Entertainment News of the Day" will be appearing "free" in both theaters, offering unusual in-depth reports on movie and theater personalities and events . . . plus trailers of HTN's present and future cinema attractions.

On this continuing show emphasis would, of course, be given to those programs scheduled for the current evening . . . to insure the largest subscriber audience possible.

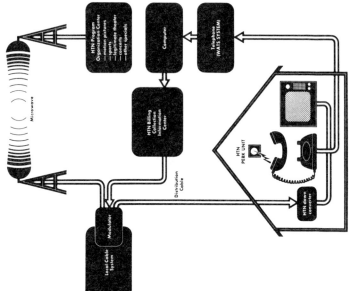

HOW THE HOME THEATER NETWORK SYSTEM WORKS

(1) Utilizing only three strategically located origination facilities and a nationwide microwave system, HTN will deliver via common carrier microwave two channels of programming to a receiving dish located at the CATV system's head-end.

(2) A modulator will convert these high frequency signals to a proper RF output and send them down the cable system's mid-band to the subscriber's home.

(3) On the back of the subscriber's TV set, HTN will attach a small down-converter which will convert these mid-band frequencies to a low VHF frequency.

(4) The subscriber can activate his down-converter by simply dialing a toll-free number on his telephone and depressing buttons on his battery-powered PERK unit, sending a coded signal to a central computer.

(5) The computer will acknowledge the request, identify the subscriber, log his order in the billing storage, and send out an appropriate response via high-speed telephone access lines to the head-end and down the cable.

(6) The subscriber's terminal unit will recognize the digitally-addressed response, open a circuit, and in seconds he will be able to view his requested program. At the program's termination, his unit will turn off automatically unless new instructions are furnished to the computer.

(7) At the end of each month the subscriber will be sent an itemized statement of his purchases and will have the option to be billed through several major credit cards.

Fig. 6-6. Home Theater Network's common carrier microwave system.

1. Utilizing only a few strategically located origination facilities and a nationwide microwave system, a quality signal will be delivered by common carrier microwave to a receiving dish located at the CATV system's head end (Fig. 6-6). That signal will be modulated onto a proper rf output and be passed, via two unused midband channels, to the subscriber's home.
2. Home Theater Network will attach a small converter on the back of the subscriber's tv set. It will convert these midband frequencies to some low vhf frequency.
3. The subscriber can activate his converter by simply dialing a number on his telephone and sending a digital message via a remote ordering unit to a central computer (Fig. 6-7).
4. The computer will acknowledge the request, identify the subscriber, log his order in the billing storage, and send out an appropriate response via high-speed telephone access lines to the head end and down the cable.
5. The subscriber terminal unit will recognize the response and open a circuit. In seconds the subscriber will be able to view his requested program. At the end of the program, the subscriber unit will turn off automatically unless new instructions are furnished to the computer.

Are additional unused channels required on the home television set to receive Home Theater Network's services?

No additional unused channels are needed. Every evening six programs will be shown on two channels (three on each channel).

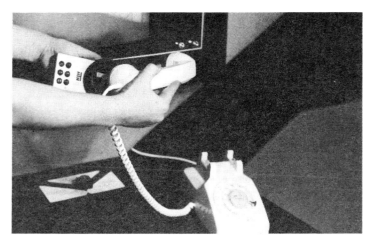

Fig. 6-7. Battery powered PERK unit sends, by telephone, a coded signal to the central computer to actuate the converter.

The subscriber may have his choice of programs by activating the converter, which is installed between the tv drop system and his set. Programs will consist of current motion pictures, rock concerts, sports events, and other specials. This will present an attractive viewing package to the potential subscriber, giving him one more reason to desire cable service.

How does Home Theater Network's pay tv concept vary from other proposed pay tv systems?

Home Theater Network claims several advantages over other pay tv systems. The company will use a nationwide microwave system to deliver two program channels to cable systems. This will eliminate the need for separate origination studios, with the additional equipment and manpower problems. Picture quality and reliability will be maximized by this single nationwide system. Transmission of live programs will also be possible. Small systems, which otherwise could not enjoy pay tv services, will be able to receive programs from the nationwide system.

The telephone ordering system used by HTN will permit impulse buying by eliminating the need to prepurchase cards or tickets outside the home. This convenience should increase program purchases; therefore, it will increase the cable operator's profit. HTN will provide a billing and collection service for the operator, and he will receive a percentage of the gross revenues.

How does the viewer know what programs are being offered on the pay tv channels?

Home Theater Network will produce programs featuring a Hollywood personality as "anchor man." These programs will immediately precede the evening's viewing fare and will incorporate previews, "outtakes," interviews, etc., to promote and identify the program fare on both channels. Additionally, the subscriber will receive continually updated program guides.

How are the pay tv channels controlled?

Each pay tv program on either of two channels can be turned on or off by digitally encoded signals from a centralized computer. The computer, on a call from the subscriber, relays instructions to the cable tv head end, allowing the selected program to be decoded on a preselected vhf viewing channel (channels 2 to 13).

How does the Home Theater Network viewer request a pay tv program?

The subscriber dials a toll-free number to contact the computer. He then indicates his viewing choice on HTN's ordering unit. He then interfaces the unit with his telephone, and the unit sends a digitally coded message to the central computer. The computer recognizes this signal and sequentially issues the instructions described in the previous question.

What programs are booked for Home Theater Network television?

Current motion pictures, live symphonies and rock concerts, Broadway plays, sports events, college accredited courses, and other specials will be offered. Additionally, HTN plans to make available a large selection of free programs. An "at home" shopping service has been incorporated in conjunction with a major catalog merchandiser. A top educational consultant has been hired to produce free educational programing. To aid cable operators in their response to FCC rules, HTN hopes to obtain a ruling from the FCC that will consider free programs to be cable system program originations.

Will Home Theater Network use X-rated films?

No.

When will Home Theater Network become available to the home CATV subscriber?

Home Theater Network expects to start operating in mid-1973 in certain areas of the U.S., with national expansion shortly thereafter.

What does a CATV operator do to carry Home Theater Network on his system?

The CATV operator leases, on a nonexclusive basis, two midband channels and narrow-band control frequencies for a period of five years. The operator has a five-year renewal option, if his system is technically able to utilize the service. The cable operator must provide space at the head end for Home Theater Network's communications equipment. This equipment is interfaced with the cable head end. Maintenance of the HTN home units is the operator's responsibility. The CATV operator must also make certain that the midband channels are carried without deteriorating the pictures.

What are Home Theater Network's responsibilities to the cable operator?

For the two leased channels, Home Theater Network will pay the cable operator 10 percent of the net amount collected from subscribers in the cable operator's systems. HTN will supervise the installation of all equipment and will arrange for all programing to the cable operator's head end.

The company will provide the two-piece home unit, which covers both conversion and ordering devices. They will market and install the services for all subscribers, and they will process and service all billing.

Does the cable operator have to buy the converter or can he lease them?

Home Theater Network furnishes its specially designed converter for its midband channels. However, should the operator require a multi-channel converter in the near future, HTN can provide a unique dual-purpose converter, providing both pay tv capability as well as regular converter usage.

OPTICAL SYSTEMS CORPORATION

What is Optical Systems Corporation?

Optical Systems is a public company run by industry veteran, Geoffrey Nathanson. The company is divided into two operations. The communications division is involved in the development of broadband communications technology and services for the CATV industry. The cinemagraphics division is developing and marketing various applications of the De Joux optical dissolve process of motion pictures projection, with particular emphasis on application of this process in the production of animation. Optical System's home offices and laboratories are located in Los Angeles, California.

How is Optical Systems involved in CATV?

Optical Systems has developed a unique technology that expands the channel capacity of a CATV subscriber's television set. An encoding/decoding process makes possible the marketing of entertainment and specialized services to cable television subscribers. This is called Private Channel Television (Optical Systems' name for pay tv).

Fig. 6-8. The Channel 100 program selector.

How does Optical propose to implement Private Channel Television?

Optical leases unused channels, from CATV system operators in the United States and Canada, to use for its Private Channel Television services. In addition, Optical will license others who purchase their proprietary equipment and market programing. They will be licensed to operate in certain selected areas or on an exclusive or semi-exclusive basis.

How does Optical create its pay tv channels?

A television picture is encoded at the source of its origination or at the CATV system's head end. The signal is sent out on a channel which has been leased from the CATV operator. This signal can be decoded by a cable subscriber if he has an Optical Systems decoder (Fig. 6-8) in his home and if he has been charged for the program which is being transmitted.

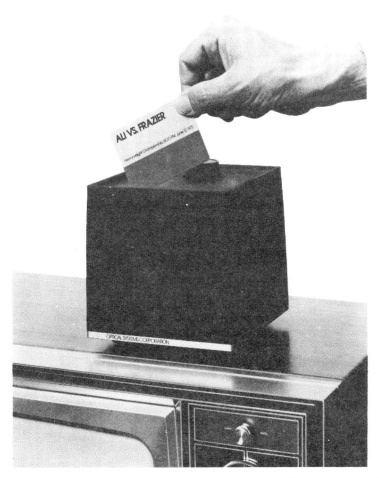

Fig. 6-9. Optical Systems' Private Channel Television concept utilizes a small "black box" which enables CATV subscribers to view programs.

What markets has Optical Systems selected for pay television and when will it begin?

Optical Systems plans to offer Private Channel Television programing to cable subscribers in San Diego, Santa Barabara, and Bakersfield, California in 1973. Optical has contracted with Trans Video Corporation, a subsidiary of Cox Cable Communications, for this purpose. The company has also contracted to lease channels from Buckeye Cablevision in Toledo, Ohio.

Optical has entered into a contract with Premier Cablevision, Canada's largest CATV operator, for the formation of a Canadian corporation which will offer pay tv services throughout Canada. Premier owns and/or operates CATV systems in Vancouver, Victoria, and Toronto. Premier management is also involved in the ownership of cable systems in Montreal, Quebec City, and other communities in eastern Canada.

San Diego, with 75 thousand potential customers, is the largest CATV system in the United States. Vancouver, with 160 thousand, is the largest in North America. In those cities where Optical will conduct its pay television business, the company will operate under the name "Channel 100."

What kind of programing will be offered to Channel 100 customers in San Diego, Vancouver, and other cities?

Current motion pictures, sporting events, and other entertainment which is generally not available on conventional television will be offered by Optical.

How will the cable subscriber purchase the right to view the program?

The decoder in the subscriber's home will be activated by the insertion of small plastic or cardboard tickets the size of a credit card (Fig. 6-9). These prepunched cards may be read photo-optically by a device within the terminal unit. Tickets will be purchased on an individual event basis or for a series of events.

What motion pictures will the subscriber be able to see, and how will they be scheduled?

Channel 100 customers will have the opportunity to view current motion pictures that have not been released to television. These will include many films that are playing, or have played recently, in local motion picture theatres. Two films will be offered each week. Normally, the films will be shown four times each day.

What will the cost be for film presentations?

The cost of tickets will vary. For example, a customer might purchase a weekly ticket for $2.25. On the other hand, he might purchase a season film ticket for as little as $1.50 a week. In this case, he would receive a plastic card entitling him to view all the films.

What kind of sports will be offered and how will they be sold?

Channel 100 will offer professional and amateur sporting events that cannot be seen on regular television. This includes the games of local sports teams which are generally "blacked out" in the area, and special events which are usually available only on a closed-circuit basis in theatres.

Tickets to these events may be purchased on an individual basis or by the season. Prices will vary from $1 per game on a season ticket basis to $5 per event for a special program, such as a heavyweight championship fight.

Where will the customer buy his tickets?

He will purchase his tickets at any ticket office, the Channel 100 service office, or by mail or phone.

Will there be any way to order a program for immediate viewing?

Yes, Optical has developed a do-it-yourself ticket which it calls the "Wild Card" (Fig. 6-10). These cards are unique in that all of the possible punch holes have been perforated and numbered. To request a particular program, the subscriber merely phones Channel 100's answering service and receives a special code which corresponds to the numbers printed on the card. By punching out the proper numbers and inserting the card into the box, he may view the program which he has requested. The Wild Card will operate his decoder for only the particular program he requests. Wild Card customers will pay a service charge for this accommodation.

How will the customer pay for his tickets?

The customer will generally purchase his tickets in person or by mail in advance. If the subscriber has a charge account with Optical, he will be able to place his order by phone, whether it be for advance sale or Wild Card tickets. He will be billed directly by Optical or through a bank credit card system.

Will a subscriber be able to use his film ticket to watch a basketball game?

No, Channel 100 will transmit a different code with each event or series of events being originated. Decoders are capable of processing 64 thousand different codes.

Will one subscriber's ticket work in his neighbor's decoder?

This is not likely. The logic circuits within the decoders have been permuted to reduce the possibilities to odds of 40 thousand to 1.

How is Channel 100 marketed?

Conventional advertising and sales methods, including newspapers, radio, direct mail, circulars, television, and direct sales will be utilized.

What kind of financial commitment must the customer make in order to subscribe to Channel 100?

The subscriber will not be required to make any commitment relative to the purchase of pay tv programing. He will pay only for those events he selects to view on his private channel.

Fig. 6-10. The Optical Systems' "Wild Card" enables the subscriber to make a last minute choice of program by punching out the appropriate numbered holes in the card.

A $20 deposit will be required, but it will be refunded when the subscriber terminates service. In addition, he will be required to purchase a service and maintenance contract which will cost $12 a year. He will receive full warranty and trouble call service at no additional charge.

What will Optical Systems' arrangement be with the cable operator?

Optical Systems will lease unused channels from the cable operator for a fixed monthly rental. This rental amount will be applied against a percentage of the gross receipts as generated by Optical through the use of those channels.

What will the cable operator's commitment be to Optical Systems?

The cable operator will not be required to make financial investment in the private channel operation. In some cases the operator may be required to install a special coaxial link or CARS microwave for the interconnection of multiple head ends.

Channel 100 will provide and install all pay tv transmitters, encoders, decoders, and origination equipment. Channel 100 operates out of its own office in the community and will generally originate programing from that location. It also will pay any cost involved in the interconnection of the Optical origination facilities with the cable system's head end.

Channel 100 will do all its own billing and collection. The cable operator will keep records on new subscribers, terminations, relocations, and information relative to subscribers who have poor payment records.

Will Optical's pay tv system require the CATV operator to install a two-way distribution system?

No, Optical's unique method was specifically designed to function on virtually all one-way CATV and MATV systems.

Is the terminal unit restricted to one channel only?

No. Optical's home terminal unit can utilize up to three standard or nonstandard vhf channels to transmit Private Channel programing. Optical also offers a special "plug-in" decoder module which interfaces with most converters, enabling the operator to make any or all channels private.

What channels are utilized for pay tv?

The system can use virtually any standard or nonstandard vhf frequency. For practical application, Optical recommends the use of nonstandard channels such as sub-band, midband and superband frequencies. These will vary with the frequency range of the cable operator's distribution system.

What if the cable operator has a 12-channel system with no available or unused channels?

Optical can utilize channels that are unused during nonduplication black out or those that are unused because of ambient air signal problems. They would, however, prefer to utilize nonstandard channels in the sub-band or midband. There are certain technical limitations in the use of these channels on most 12-channel systems; the number of usable interference-free nonstandard channels is usually limited.

SKIATRON ELECTRONICS AND TELEVISION CORPORATION

What is Skiatron Electronics & Television Corporation?

Skiatron is a public company headed by Arthur Levey, a television pioneer. Skiatron, insofar as pay tv is concerned, has demonstrated systems suitable for both over-the-air operations and closed-circuit cable transmission. Either may be integrated for use in CATV systems.

In the 1950s, Skiatron conducted tests over WOR-TV (New York City) for approximately eight years, with FCC authorization. The Skiatron Subscriber-VisioN cable system was utilized for large-scale commercial operations in Los Angeles and San Francisco. For nearly six months the system transmitted programs in both color and monochrome to 7000 subscribers. This programing was halted by a referendum. The referendum was subsequently declared unconstitutional by the California Supreme Court and upheld by the United States Supreme Court.

What is Skiatron's special area of activity?

Skiatron has a strong patent position in the field of cable television with centralized metering. The system has two-way capability and should be adaptable to the new specifications of the FCC.

What is the nature of Skiatron's services?

The Skiatron system is designed for large urban communities such as Los Angeles and San Francisco. The system performed successfully in those cities and had the capacity to expand to an unlimited number of subscribers.

Are additional channels required on the home television set to receive Skiatron's services?

Yes, three or four midband channels could be used. System channels are available via the Skiatron program selector. The program selector is attached between the cable outlet and the antenna input leads of the subscriber's tv set.

How does Skiatron's pay tv concept vary from other proposed pay tv systems?

Patented principles of centralized billing (Fig. 6-11) are unique in the Skiatron system. The system also has two-way communications capability (nonvoice), proven in large-scale commercial operations.

How does the viewer know what programs are being offered on the pay tv channel?

Subscribers will receive program guides, listing the details and the cost of each program. The pay tv channel will also offer previews or "sample periods" for its programs. One channel will be used specifically to give the subscriber information on the program fare. (At times, the information channel may be used for programing; thus making four program channels available.)

How are the pay tv channels controlled?

At the end of the sample period, the central computer will automatically check each set to determine which channel is in operation. The computer then records name and charge for subscribers who continue to view after the sample period. Billing will be monthly, by mail. No coins, meters, tapes, or servicemen are necessary. Since channels are controlled via the computer at the Skiatron headquarters, a computer readout is available at any time.

How does the Skiatron viewer request a pay tv program?

He selects the program on the Subscriber-VisioN control unit. When he chooses the correct channel, the decoder becomes operational.

Fig. 6-11. Arthur Levey, president of Skiatron, operating a central billing system capable of handling 450 thousand closed-circuit subscribers.

What programs are booked for the Skiatron television service?

Skiatron has only recently regained its exclusive U.S. license and is now examining program possibilities. Quality programs from major sources are sought. Possible sources include current films, legitimate theatre (both Broadway and off-Broadway), British programs and the sports world.

Will Skiatron use X-rated films?

No, only films considered suitable entertainment for the entire family will be made available.

What does the CATV operator do to carry Skiatron on his system?

Skiatron Subscriber-VisioN franchises are available to CATV operators. The franchises enable operators to integrate subscriber television into their CATV systems. The franchise operator obtains the program selectors required for his subscribers from Skiatron.

What are Skiatron's responsibilities to the cable operator?

Skiatron is responsible for the delivery and maintenance of a reliable service to the home subscriber.

Does the cable operator have to buy the converters or can he lease them?

Present plans give the cable operator the option to either purchase or lease the equipment.

TELEPROMPTER CORPORATION

What is TelePrompTer Corporation?

TelePrompTer is the largest cable television company and the industry's leader in development of nonbroadcast cable communications services.

TelePrompTer has approximately 12 thousand shareholders and assets over $240 million. The company entered the CATV field in November 1959, and continues to expand its commitment to community-responsive programing, research, and development. Subscribers are assured of the highest quality and broadest range of services made possible by cable technology.

What is TelePrompTer's special area of activity?

About 73 percent of the company's revenues were generated by cable tv in 1972, primarily from monthly fees for basic services. Tele-PrompTer cable tv serves 760 thousand subscribers (as of February 15, 1973) with basic services of improved reception and greater program diversity. The company operates 138 cable systems serving communities in 33 states and two Canadian provinces. TelePrompTer is the only cable tv company that submits an Annual Report on Program Origination to the Federal Communications Commission. As of January 1973, 95 systems originate local programing.

When the wired nation is finally achieved, it is expected that new, nonbroadcast services will become more numerous and more important to the community than basic services. Many of these new services are already under development and testing by the company; others are still in the research stage.

Listed below are some of the many services which are likely applications of broadband cable technology:

Pay television programing
Shopping at home
Fire and burglary alarms
Surveys and polls
Delivery of health care information
Utility meter reading
Fares and ticket reservations
Computer-aided educational instruction
Traffic control
Newspaper and reference material in home printouts
Message and mail delivery
Payment of bills and transfer of bank funds
Educational opportunity applications

At this time it is impossible to predict when these services will be provided on a commercial basis. Some, such as merchandising, security alarms, and telemedicine services, are already in the developmental stage. Still others, such as pay television offerings, motion pictures, and sports programs, are expected to be commercially introduced in 1973.

What is the nature of TelePrompTer's services?

One of TelePrompTer's most recent advances is the development of a new "Premium Television" (or pay tv) system, designed by the Magnavox Company to meet specifications supplied by TelePromp-

Ter. During a cooperative feasibility study conducted in 1972, the equipment was jointly demonstrated to the press and the financial community. The system was demonstrated to the cable, sports, and performing arts industries on February 1, 1973.

The new equipment permits cable television subscribers to instantaneously select films, sporting events, and other special-interest programing on a per-program basis. Telephone calls, advance ticket purchases, plastic cards, and other devices required by competing systems are not necessary.

The system, designed for use on existing one-way cable television operations, permits cable subscribers to preview pay tv programs before making a selection. The viewer selects programs by pressing an acceptance button on his home terminal. This instantly activates a control unit which unscrambles the program and simultaneously records the purchase for computer processing of the customer's bill. Data readouts are periodically collected from control units, which can store information for up to 32 homes.

Are additional unused channels required on the home television set to receive TelePrompTer's service?

TelePrompTer's services are all compatible with any standard home television set. The company provides all needed equipment for basic cable tv service and for pay television and other nonbroadcast services. The cable converter expands over-the-air channel reception to 26, 30, or more channels.

The system transmits a scrambled sound and picture on a nonstandard channel of the television set. The company provides the unscrambling device that allows program selection and brief free previewing of programs.

How does TelePrompTer's pay tv concept vary from other proposed pay tv systems?

The TelePrompTer/Magnavox Premium Television System (Fig. 6-12) makes possible impulse selection and purchase of individual programs, with no inconvenience to the subscriber. By providing a single ACCEPT button, the tv process eliminates the need for phone calls, ticket purchases, plastic cards, etc. The system also provides highly efficient operation for CATV operators, automatically recording the date, time, place, name, and price of a program for billing. The equipment is designed for use with two-way, broad-band cable operations, thus eliminating the possibility of obsolescence.

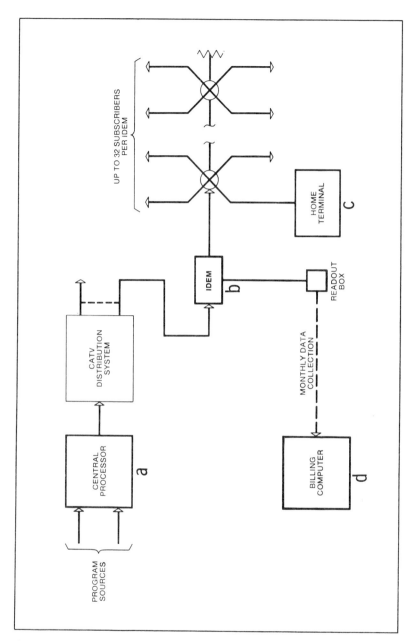

Fig. 6-12. Schematic diagram of a standard one-way cable tv system with four major Premium Television units added: (a) central processor; (b) Interactive Data Exchange Module (IDEM) and readout terminal; (c) home terminal; (d) billing computer.

How will the viewer know what programs are being offered on pay tv channels?

Pay tv programs will be promoted in newspapers and in program guides mailed monthly to subscribers. Daily program schedules will be listed on the guide channel, a message wheel channel, and the time and weather channel. Subscribers will also be informed about offerings during breaks in local origination programing.

How are the pay tv channels controlled?

The channels are controlled by the home terminal and Interactive Data Exchange Module (IDEM), the heart of the new system (Fig. 6-13). Located near subscriber homes, the IDEM's primary functions are to: (1) receive and process head-end program commands; (2) receive pay tv requests from as many as 32 subscriber terminals in its domain; (3) control the unscrambling circuits within these terminals; and (4) act as a neighborhood "box office" by collecting and recording data for billing purposes. Data readout of home charges are collected periodically and fed into a control computer for billing (Fig. 6-14).

For the subscriber's protection, this system provides a distinct control over each viewing channel. Each channel selection button has a corresponding keylock. This feature permits the subscriber to limit children's viewing selections and prevent unauthorized purchase of pay tv programs.

How will the TelePrompTer viewer request a pay tv program?

After the preview period, the subscriber will push the ACCEPT button on his home terminal, unscrambling the program of his choice (Fig. 6-15).

What programs are booked for TelePrompTer?

First-run motion pictures, live sports events, Broadway shows, nightclub and cabaret entertainment, opera, concerts, and other major events will be offered to subscribers.

Will TelePrompTer use X-rated films?

The use of X-rated films and its ramifications are currently under study. The company plans to offer a wide range of program choices to meet the diverse interests of the public. Programs will be consistent

Fig. 6-13. The Interactive Data Exchange Module or IDEM top) is at the heart of the system. Data readout of home charges are collected periodically (bottom) and fed into a central computer for billing.

Fig. 6-14. The solid-state central processor (right) controls the operation of the Premium program origination sources and generates commands to an IDEM through standard CATV head-end facilities (left).

with good taste, good judgment, and accepted standards of ethics and moral behavior in the community.

When will TelePrompTer become available to the home CATV subscriber?

Field testing of Preminum Television will be conducted in a Tele-PrompTer system in late spring. A larger marketing and engineering test is scheduled later in 1973, and full production of the equipment by Magnavox is targeted for the end of 1974.

Fig. 6-15. The viewer may preview a pay tv program by pressing a channel control on a special home terminal unit.

What does a CATV operator do to carry Premium Television on his system?

The hardware for this pay tv system is manufactured and distributed by the Magnavox Company and is available to call CATV system operators. Premium programing produced by TelePrompTer for its systems is also available, on a negotiated contract basis, to other cable systems that have pay tv capability. Prospective pay cable operators should contact:

K. L. Hutchinson
Vice President, Marketing
Magnavox Government and Industrial Group
1700 Magnavox Way
Fort Wayne, Indiana 46804

THEATREVISION INCORPORATED

What is TheatreVisioN Incorporated?

TheatreVisioN is an affiliated company and a joint venture of Chromalloy American Corporation and Laser Link Corporation. Laser Link created and developed the TheatreVisioN systems of pay tv. Worldwide patents have been issued to the company for this system.

The TheatreVisioN pay tv system is the modern counterpart of the box office found at conventional motion picture theatres, sports arenas, and concert halls. The device used by TheatreVisioN is a single mechanical-electronic device that attaches externally to the terminals of any tv receiver. It provides a variety of programs for customers whose homes have been linked to CATV systems.

How does TheatreVisioN serve the cable industry?

TheatreVisioN has developed a multichannel capability in a compact decoder device. The decoder can provide three or more simultaneous television programs in either midband, superband, or sub-band television frequencies.

TheatreVisioN has begun to lease CATV channels, presently in the midband frequencies, from cable operators for the provision of current films and other forms of television entertainment. Its key element is the electronic ticket.

What are electronic tickets?

Laser Link Corporation, which owns 45 percent of TheatreVisioN, has created a patented method of impinging thousands of various codes on an electrostatically encoded plastic ticket. This enables the company to offer current motion pictures and other events to CATV subscribers. The method is based on the historic pattern of selling entertainment to the public through the use of a ticket and a box office (Fig. 6-16).

What is the relationship between the cable operator and Theatre-VisioN?

TheatreVisioN is an exhibitor of all forms of box office entertainment and serves cable television systems in all areas. It provides a total service to its affiliated cable operators by leasing channel space from the operators' systems.

Where is TheatreVisioN operating today?

On December 1, 1972, a pilot TheatreVisioN system was inaugurated in conjunction with Storer Cable Television, a multiple systems CATV operator and a division of Storer Broadcasting Corporation. The pilot system, located in Sarasota, Florida, services one thousand subscribers of the 20 thousand home cable system operated by Storer. In addition, TVN is available to hundreds of rooms in hotels and motels in Sarasota.

What is the difference between TheatreVisioN's CATV decoder and the motel decoder which was designed for Philco-Ford?

They differ in operational criteria as follows:

	Motel Decoder	CATV Decoder
Number of different entertainment channels	2	3
Required number of exclusive codes per channel	8	256†
Acceptance of TVN precoded ticket	yes	yes
Allowance for ticket destroy transmission from remote location*	yes	yes
Capacity to store minimum of 100 destroyed tickets	yes	yes
Allowance for multiple programing of different codes on one channel in one night	yes	yes
Internal destroy "clock" actuated within 8½ to 15 hours*	no	yes
Functions with fully scrambled TVN channels	no	yes
Facility for simple, secure ticket emptying	yes	no

* Must be functional in both "TheatreVisioN" and "Local-TV" switch positions.

† Second generation has 1,024 codes.

How many channels of service are being offered in Sarasota?

TheatreVisioN has leased three channels in the midband (G, H and I) on the Storer Cable.

What type of programing does it provide?

The majority of programs available are motion pictures provided by the top Hollywood studios. Mr. Dore Schary, president of Theatre-VisioN, has developed a format of offering seven motion pictures

Fig. 6-16. The author (left) pointing to an encoded ticket taken off a seven ticket card held by Dore Schary, chief executive officer of TheatreVisioN. James Hall, vice president of Storer Cable Television is in the center.

monthly. There are three films presented nightly on even dates; three other films are transmitted on odd dates. The remaining film, a special attraction, is scheduled on the weekends.

How does a subscriber purchase a program?

First, he must request the installation of a TheatreVisioN decoder. The decoder has a slot for electronically coded tickets. A switch in the decoder enables the subscriber to watch either TVN or local tv.

This unit is installed without charge to the subscriber, although he is required to purchase the first month's programing of seven attractions for $14. Once it is installed, the subscriber has access to the multiplicity of programs available.

To see a specific program, he inserts the program ticket into the slot. Within the unit, the ticket is electronically read and produces the program purchased. All programs feed through a single channel of the television set. The film is shown four times nightly, and the viewer has the option of watching one or all showings. He may switch back to local television or shut off the set and return to Theatre-VisioN programing later.

What do TheatreVisioN programs cost?

The entire family can watch TheatreVisioN programing for a total of $2 per film. This restores a popular price which has disappeared from conventional theatres in recent years.

How about impulse buying with electronic tickets?

After using TheatreVisioN for one month, the subscriber has the choice to purchase individual tickets or a strip of tickets for all films in the month. If the subscriber purchases the strip, he may use as many of the tickets as he wishes and return the remainder for credit. It is possible for the subscriber to buy on impulse if he has the month's supply of tickets on hand.

How does TheatreVisioN interface with a cable system?

When TheatreVisioN enters into a lease agreement with a cable operator, it provides the following: all decoders, the encoding unit, electronic tickets, service and maintenance of decoders, promotional material, and films for the system.

The encoding equipment feeds TheatreVisioN programs into the midband of the cable system. Programs are then transmitted by the cable system to the subscribers.

What kind of investment is required by the cable operator to carry TheatreVisioN's service?

The cable operator provides video cassette equipment in either manual or automatic form. Under present prices of manual video cassette equipment, this would cost the cable operator less than $10 thousand in capital investment.

How does TheatreVisioN differ from other forms of pay tv?

Since TheatreVisioN does not require computerization or two-way cable capability, it can be integrated, at low cost, into any cable system. TheatreVisioN was one of the few pay tv systems in the U.S. offering service in the home at the start of 1973. Other systems requiring computer interconnection, two-way capability, and telephone interconnection were operating in hotels in various parts of the country. None serviced the home audience using a per-program charge.

TheatreVisioN uses ticket dispensers of the type shown in Fig. 6-17. These are well distributed within the area to provide easy access

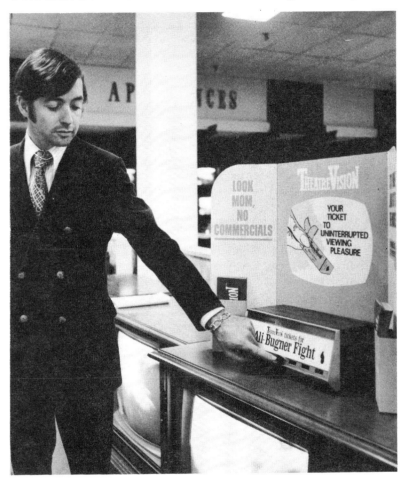

Fig. 6-17. TheatreVisioN unit installed in department store to sell tickets for the Ali-Bugner fight telecast.

to subscribers. Many times special sporting events are available on short notice. This is due to the time needed to negotiate agreements when telephone company microwave links must be used. When a program becomes available on short notice, local advertising will advise subscribers to visit the ticket dispensers.

Where does TheatreVisioN plan to go from Sarasota?

During the next three years, TheatreVisioN plans to expand its system throughout Florida. Negotiations are now being completed with cable operators in major cities of the United States and Canada.

TRANS-WORLD COMMUNICATIONS

What is Trans-World Communications?

Trans-World Communications is a division of Columbia Pictures Industries, Incorporated, a major supplier of entertainment and communication services. Columbia is also in television and radio broadcasting, television production and distribution, video tape production, and other services.

What is Trans-World's special area of activity?

Since 1968, Trans-World has devoted itself to producing various forms of closed-circuit communications and entertainment services for use in hotels throughout the U.S. and Canada (Fig. 6-18).

What is the nature of Trans-World's services?

Trans-World offers a Tele/Ad service on a vhf channel over closed-circuit tv in hotels (Fig. 6-19). Tele/Ad is a program that informs tourists of current attractions in the city they are visting. The program can be viewed at all times. It is currently available at hotels in the United States, Canada, and the United Kingdom.

A Tele/Vention service is used for closed-circuit coverage of conventions. This service is broadcast over a previously unused vhf channel. Tele/Vention was used for coverage of both the Democratic and Republican National Conventions in July 1972. Coverage was provided for over fifteen thousand rooms in several Miami hotels. Two years in a row Tele/Vention provided coverage at the five-day National Association of Home Builder's Convention in Houston.

Tele/Theatre is another Trans-World service. Tele/Theatre provides three channels of pay tv entertainment, including motion pictures, live sports events, and other special programs.

Are additional unused channels required on the room television set to receive the Tele/Theatre service?

No. A Program Selector Box (Fig. 6-20) is attached to each television receiver. This unit adds four additional controllable channels in the midband frequencies.

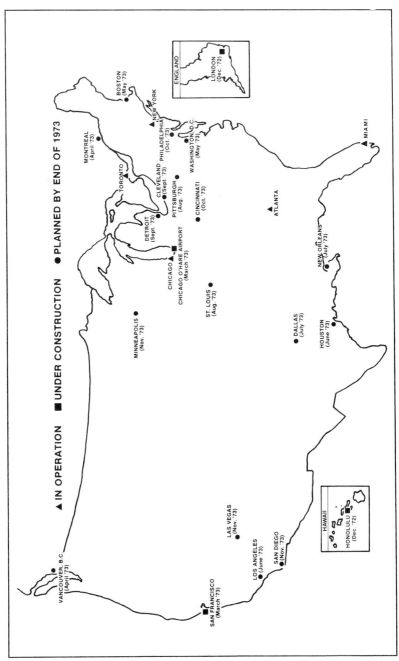

Fig. 6-18. Trans-World studio origination network.

How are Tele/Theatre programs transmitted to the hotel guest room?

Programs are transmitted from a central origination studio (Fig. 6-21) in each city. The studio is equipped with an origination console built by the International Video Corporation, utilizing a one-inch video format.

Are the signals broadcast over the airways?

No. They are fed by a closed-circuit coaxial cable to each hotel and then distributed over the hotel's master antenna system into each guest room.

How does the Trans-World marketing concept vary from other proposed pay tv systems?

In the Trans-World system, the viewer pays for each separate program or event he chooses. The Trans-World System avoids the bother of advance ticket purchase and does not use special plastic control cards, or other paraphernalia.

How does the viewer know what programs are being offered over the pay channels?

Tele/Theatre has one channel specifically for previews. It runs 24 hours a day at no charge to the viewer. To see these free previews, the viewer switches his Program Selector Box to the preview channel.

How are pay channels controlled?

Each pay channel is selectively turned on or off from the head end (origination studio) via digital control signals that are transmitted down the coaxial cable. Each Program Selector Box contains a "read only memory" module that has been preaddressed with a unique code. It will respond only when command by the Monitor Central Computer (Fig. 6-22) or by a manually operated command generator (Fig. 6-23). The hotel program selectors and digital control systems are built by K'SON Corporation.

How does the Tele/Theatre viewer request a pay program?

He simply dials a predesignated telephone number.

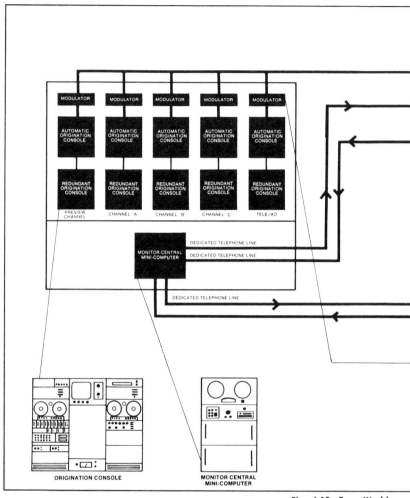

Fig. 6-19. Trans-World

Why the telephone?

There are only three ways the viewer can communicate with the program source, and the telephone best suits the immediacy of Tele/Theatre. The telephone is fast and easily accessible to the viewer. The other two methods of communication, two-way cable or mail, are either not currently feasible or inadequate. Two-way cable capability would return an electronic order from the viewer; this method is not presently feasible. Ordering programs by mail does not accommodate the impulse buyer.

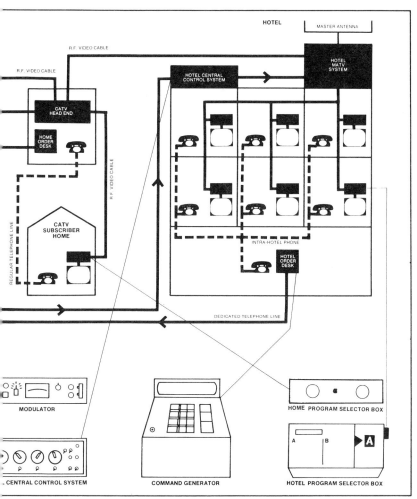

origination studio.

Does Trans-World only program Columbia motion pictures on the Tele/Theatre network?

No. Since the introduction of Tele/Theatre in December of 1971, Trans-World has booked motion pictures from the seven major distribution companies, plus many independents.

What films are booked for Tele/Theatre?

The best available films will be shown. Tele/Theatre will not book X-rated films.

Fig. 6-20. Trans-World Communications Hotel Tele/Theatre Program Selection Box for use in hotel/motel installations only.

Will Tele/Theatre be made available to the home CATV subscriber?

Yes. The first 10 thousand Program Selector Boxes (Fig. 6-24) for CATV home use will be manufactured by Oak beginning in early 1973.

Fig. 6-21. The originating studio of Trans-World Communications' closed-circuit television system in the Columbia Pictures Building, New York City.

Fig. 6-22. Trans-World Communications' Monitor
Central Mini-Computer.

What does the CATV operator do to carry Tele/Theatre on his system?

Trans-World offers the CATV operator a contract to obtain programing from one of the Tele/Theatre origination studios. The operator must install Oak/Tele/Theatre modified Gamut 26 converters in the homes of his subscribers.

What are Trans-World's responsibilities to the cable operator?

Trans-World will supply a preaddressed K'SON plug-in module for insertion into the subscribers' converter. The plug-in will control the subscribers' pay channel and will interface with the Trans-World computer for billing. Trans-World will program and operate the

Fig. 6-23. Trans-World Communications' Command Generator.

Fig. 6-24. Trans-World Communications' Home CATV Tele/Theatre Program Selection Box.

origination studio for a minimum of 18 hours a day, seven days a week. Trans-World will supply the CATV operator with a computer readout of all the subscribers' orders, either weekly or monthly. For CATV operators who have their own computer for billing, Trans-World will supply an IBM compatible magnetic computer tape. This will enable the operator to automatically invoice his subscribers.

Does the CATV operator have to buy the converters or can he lease them?

He can lease converters on a five-year lease/purchase plan.

Does the Oak/Tele/Theatre converter have special features?

Yes. In addition to the four controllable pay channels, it provides 22 off-air channels and has two-way capability. By adding another module, the CATV operator with a two-way cable system can retrieve four channels of information from the subscriber's home.

How much will the Oak/Tele/Theatre converter cost?

The cost of the converter will be approximately $65, if the CATV operator decides to buy rather than lease one.

JERROLD ELECTRONICS CORPORATION

What is Jerrold Electronics Corporation?

Jerrold Electronics, a subsidiary of General Instrument Company, is the largest supplier of equipment and construction services to the CATV industry.

How is Jerrold related to pay tv?

Jerrold's head-end equipment and cable distribution networks form the "electronic pipelines" through which pay tv programs are fed. The company is developing a system called Communicom. This system will permit the home viewer to preview, order, and receive first-run motion pictures, sports events, and other special programs.

What is Communicom?

Communicom is a two-way, computer-controlled, communication system. The viewer is linked through a Home Control Center to a computer located at the CATV operator's headquarters (Fig. 6-25).

Fig. 6-25. Jerrold Electronics Corporation's Communicom
head-end computer.

Each home terminal has a unique address and is polled constantly by the computer for service requests. Along with the normal pay tv programs, the system will provide other convenient services that are not yet available.

Exactly how does Communicom work for a pay tv program?

The system will provide the viewer with a convenient and foolproof method of pay tv program billing.

Any number of channels can be programed for pay tv purpose. The CATV operator will decide the time and number of available channels. One channel will be used continuously by the CATV operator as a free barker channel. The barker channel will indicate the current pay program in progress and preview coming events, giving channel, time, and cost.

When the viewer has selected a channel which has a pay tv program in progress, a status-indicating light will be turned on at the Home

Control Center. The viewer will be allowed a short preview period before he is charged for the program. Shortly before the expiration of the free time, the status light will blink at the subscriber's home. If the viewer decides to purchase the program, he will turn an authorization key. The status light will stay on and he will watch the program without interruption. The head-end computer will record his purchase for billing purposes. If no decision is made after the expiration of free time, the program will be automatically turned off, but the status light will continue to blink. He may still purchase the program at any time with the authorization key.

With such a convenient system, couldn't the customer run up a bill beyond his intent?

Communicom offers two degrees of protection to the consumer against inadvertent high charges. Only authorized persons will be able to use the key; children cannot accidentally charge a program. Each viewer can prescribe to the CATV operator what his maximum monthly credit should be. Any purchase request beyond his chosen limit will not be granted.

How does the Communicom system differ from other pay tv schemes?

Many other systems attempt to adapt their pay tv approaches to the public's motion picture habits. Since the number of tv viewers far surpasses that of the film viewers, the Communicom approach will have a higher probability of sucecss. The subscriber does his viewing in the convenience of his home without any prearranged requirements such as tickets, coded cards, telephone, and various other schemes. In addition, the system, because of the versatile capability of the Home Control Center, provides the CATV system operator with total flexibility in methods of billing, channel usage, and program times. The system can bring the maximum participation of the tv audience for the pay tv programs.

Will Jerrold provide motion picture programs and special events to the operators?

No. As pay tv reaches a broad acceptance level, software exhibition and distribution will become a separate service from hardware delivery. This follows the current film industry practice of separation between the producers and the exhibitors.

Can the system be adapted for a flat-rate subscription pay tv service?

Yes. The flat-rate subscription scheme will be simply a sub-set of the per-program billing; the computer can implement this with a simple command operation.

Can Communicom accommodate other nonentertainment pay tv schemes?

Yes. A large market exists for the areas of adult education and enrichment courses. People can subscribe to specialized enrichment or "how to" type of programs at home, without the inconvenience of attending classes in remote areas. This will be convenient and will bring the best quality of instructional materials to the home. With the use of the computer interactive capability of the Communicom system, tv correspondence courses with attendance records, quizzes, and grade reports can be conducted without any additional hardware. This can be used both for nonaccredited enrichment courses as well as accredited courses for high school, community college, and even college-level courses. Continuing educational programs for professional trade personnel can be accommodated very easily by the system.

Is there a sufficient market for these nonentertainment-type applications?

Yes. There is an ever increasing demand by the public for more information of all kinds. People with home responsibilities, or without transportation, will be able to continue their education at home.

What is the timing for the implementation of Communicom systems in the field?

In conjunction with multiple systems operators, Jerrold will conduct major field trials of two-way Communicom systems near the end of 1973. These field trials will vary from 500 to 2000 terminals per system. The trial will probably run for a period of one to two years. Major implementation of two-way systems using Communicom will take place after the trial.

If Communicom cannot reach mass audiences until one to two years from now, what will be done in the meantime to satisfy pay tv requirements?

The company will provide a scrambling/descrambling (SDS) system to be connected to one-way systems through either the Local Func-

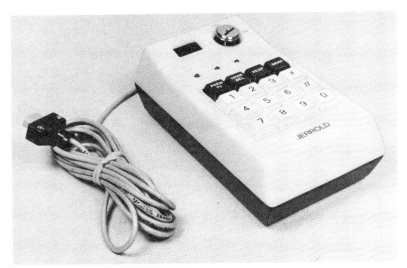

Fig. 6-26. Jerrold Home Control Center XLT, which can be upgraded in the future with a two-way module.

tioning Unit (XLT) of the Home Control Center (Fig. 6-26), or Jerrold's Standard Converter Model RSC. Neither unit requires any computer interaction. It will provide pay tv on a flat-rate subscription basis.

Why is a scrambling/descrambling system needed?

Without it, the system can be easily defeated by the installation of another standard converter.

When will the XLT be available?

Plans are to introduce the XLT, on a production basis, during the second half of 1973.

What does the XLT of the Communicom system have over standard converters?

Besides the future capability with a two-way module, XLT converters will be able to step channel advance without keyboard entry. A descrambler system may easily be added onto the converter. Some of the other features of the XLT converter are: digital keyboard entry for channel selection; independent channel display; selected channel disabling for pay tv; keylock authorization switch; remote on and off switch; and afc converter.

What is the cost of an XLT terminal?

The cost is projected to be $65, in production quantities.

What does the complete two-way Home Control Center terminal cost?

The complete two-way terminal is projected to cost approximately $150, in production quantities.

Does the two-way Home Control Center offer any additional services?

In addition to pay tv, it will offer the services of home merchandising, education and training, security systems, opinion polling, two-way message transmission, and many other two-way services. The center provides the CATV operators with methods of terminal diagnostics, centralized billing, viewing statistics for remote connect/disconnect, segmented programing, and system monitor and control.

What are the advantages and the disadvantages of using the telephone as the return channel, versus using the cable for the return information?

For pay tv to operate on a very large scale, the use of the telephone as the return channel is not efficient. The telephone network was designed for efficient voice communications between individuals, not for many subscribers to one central point. The cable distribution network is a very efficient means of two-way communication between a central point to a mass audience. In addition, the system can transmit data without human interface. It is anticipated that the Communicom system will be able to handle 10 thousand subscribers simultaneously on a two-second response time. During the course of several minutes of free viewing time, no delay will occur to service even in the most densely populated urban communities.

Will Communicom show X-rated films?

The Communication system is a delivery system and will not discern what type of rating films have. This is strictly the decision of the CATV operator.

Would the Communicom system of pay tv work in a hotel/motel environment?

Since most large hotels have a room capacity between 500 and 2000, the system would operate on a reduced basis very effectively. Using the one-way address capability of the Communicom system, the hotel could become a slave control station whereby an operator at the front desk could access any room at his choice. The return capability would be via the hotel's own internal telephone network.

Why not a two-way system in a hotel/motel application?

Most of the MATV distribution systems that exist in hotels are costly to outfit with two-way transmission capability.

Does the CATV operator have to buy the Communicom system or can he lease them?

Various arrangements of financing can be arranged with Jerrold.

CONCLUSION—CABLE PAY TV SYSTEMS
TWO-WAY SYSTEMS WITH PAY TV POTENTIAL

Who else is designing two-way cable systems with pay tv potential?

The following companies are designing systems: EIE, Incorporated, North Hollywood; Theta-Com, Los Angeles; Scientific Atlanta, Atlanta; Tocom, Dallas; Vicom Manufacturing, Ann Arbor; and Video Information Systems, New York City.

What are EIE's methods?

The home subscriber terminal has a "turn-key" operation for pay tv. By turning the key, the subscriber opens up a pay tv channel which is paid for by the amount of use. The terminal also has a four-button box (Yes, No, Maybe, Go buttons). The buttons connect to a computer at the head end for services which include merchandising, security, and subscriber opinion surveys.

Where is EIE's system being tested?

EIE is presently conducting its test system in Orlando, Florida via ATC cable. The initial test involved 25 interactive terminals.

What are Theta-Com's methods?

The subscriber turns to the pay tv channel, puts a key into the slot (free program preview if available) and pushes a button (Fig. 6-27). He is billed automatically by a computer. The subscribers can switch

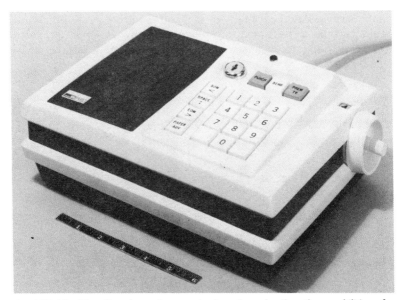

Fig. 6-27. The Theta-Com home instrument gives the subscriber the capabilities of a cable system in addition to Premium or pay tv.

to regular television and back to pay tv if they desire. Other subscriber response applications are available over eight channels.

Where is the Theta-Com system being tested?

Theta-Com's subscriber response system is being tested in El Segundo, California. The system is a Theta Cable (Hughes/TelePrompTer joint venture) project.

What are Scientific Atlanta's methods?

Scientific Atlanta offers four channels for pay tv. The viewer turns to an appropriate channel and turns on a keylock. There is a free viewing time available before billing starts. Billing could be on a per-program basis or on a time basis.

Where is Scientific Atlanta's system being tested?

Since May 1972, Scientific Atlanta has been conducting a test in Carpentersville, Illinois. The test is for "interactive tv," using LVO's cable systems.

Fig. 6-28. Tocom's home instrument includes the 26-channel converter capable of feeding signals back from the Tocom Studio.

What are Tocom's methods?

Tocom uses a computer-controlled system. There are four push buttons on the 26-channel home terminal (Fig. 6-28). The buttons are for subscriber responses; there is a switch for pay tv. The subscriber turns on his set to the pay tv channels, turns on the terminal, and puts a key into place; the program comes on, and the computer billing cycle commences (Fig. 6-29). Billing can be done on a per-program basis or a time basis.

Fig. 6-29. Tocom Studio has a complete computer console and is fully responsive to all the two-way potential of cable television.

Where is Tocom's system being tested?

Tocom is presently testing its two-way system in-house and will install it at the cable system in Irving, Texas.

What are Vicom's methods?

Vicom's system includes a subscriber terminal, a touch tone device, and a key operation for pay tv. The subscriber turns to a pay channel and pushes a button to get the program. The system allows for interactive audio and video, merchandising, data retrieval, and pay tv services. Payment for pay tv is on a per-viewing basis.

Where is Vicom's system being tested?

The company has had a system of interactive video in operation since 1971, via Telecable in Overland Park, Kansas. Sixteen terminals have been in use for educational purposes. Vicom is planning another installation in Ann Arbor, their home base.

What are Video Information System's methods?

Each VIS home terminal contains six push-button switches: four for data, one for cancellation, and one for end of message. The computer stores and displays information. The two-way terminal will be used for pay tv and services such as home shopping, alarm reporting, and CATV system monitoring. The viewer simply turns to the channels devoted to pay tv and is billed according to what he views. They plan a flat service charge per month for the terminal plus a per-program charge for pay tv. The VIS subscriber terminal has a 26-channel capacity.

Where is Video Information Systems being tested?

VIS has tested its two-way system periodically over Sterling Communications in New York City. VIS is currently building a CATV system in South Orange, New Jersey, and expects to install between 2000 and 4000 interactive terminals in 1973.

Who has developed automated video cassette player systems for cable originations?

Telemation, of Salt Lake City, has developed automated systems which are in use by TheatreVisioN in Sarasota, Florida.

Goldmark Communications Corporation, a subsidiary of Warner Communications, has developed the Star-Pak video cassette player.

Are these systems fully automatic or is manpower required after the cassettes are fed with cartridges?

The TheatreVisioN system is fully automated. However, Theatre-VisioN maintains a stand-by operator, even though it has duplicate tapes on each channel and in some instances a third safety cassette player. The system is designed to run three motion pictures simultaneously, paid for on a program basis using tickets for decoding.

The Goldmark system is designed to provide programing for a cable system. It uses four video cassette players to feed full-length motion pictures and other programs. The program messages are interspersed through automatically controlled preset timing. The Goldmark system is designed for totally automatic operation. They plan to automate systems owned and operated by Television Communications Corporation, another Warner subsidiary.

Do these low-cost video tape players have any of the distortion problems identified with low-cost video tape playback equipment?

Both TheatreVisioN and Star-Pak have installed the Goldmark solid-state skew correction system. This system prevents picture distortions caused by variations in tension on video tape as it moves through the video playback equipment. Low-cost video tape equipment was not accepted in the past because of the picture distortion. Now that the Goldmark system has corrected this disturbance, low-cost equipment has become quite reliable.

Chapter 7

Auxiliary Services

What auxiliary services will help cable pay tv grow?

The use of the pay tv systems for educational programing will help pay tv to grow. One educational system has been proposed by Dr. Alexander Schure, president of the New York Institute of Technology. Dr. Schure purposes that off-campus courses be given on especially reserved midband channels, using presently available program material. At this time, NYIT believes that courses needed for a masters degree could be taken by cable pay tv subscribers at home. These courses could only be seen by those subscribers who had received the proper decoding information (ticket, etc.) after paying their tuition to the college.

Surveys have shown that many people do not advance their education because of the inconvenience of driving to and from the school. Night courses are unpopular at some urban schools because many people do not want to go out at night.

Dr. Schure believes that advanced education should not have to be limited to "brick and mortar." Educational pay tv, as an auxiliary service to entertainment pay tv, is one of the answers to the nation's educational problems.

Are there any other educational systems which are being offered for pay cable television?

Yes, Educating Systems (of New York City) offers programmed instruction over unused fm radio channels. These courses can only be received by a special fm receiver with subcarrier push buttons. The programmed learning is placed on the special fm carrier by a Sylvania-designed subcarrier modulator. These subcarriers are received when the fm radios' push buttons are depressed. Educasting Systems plans to market these radios through CATV operators.

Are any of the pay tv proponents enthusiastic about the potential income from auxiliary services?

Yes. Home Theater Network and the University of Southern California have joined in an effort to program a Bachelor of Arts degree via pay cable. Anyone who participates will be able to "attend" classes at USC. The university will not indicate, on transcripts or on the degree, that it was a special or correspondence course.

Home Theatre Network is also working with catalog companies to present computer purchasing of merchandise on the pay tv system. Plans are being made with major record companies to arrange for sample recordings to be transmitted in stereo to the home. Records may then be purchased from the home.

What auxiliary services have been developed by TelePrompTer to offer income to the cable system?

Some of the new services are being developed by TelePrompTer's subsidiaries.

Muzak Corporation arranges, records, and distributes functional music programs designed to relax or stimulate people at work and at leisure. An estimated 65 million people around the world hear Muzak programs every day.

Filmation Associates, well known for its animated cartoon productions, produces regular and animated programs for both broadcast and cable television.

TPT Communications, a communications contracting company, provides such services as private business telephone and data systems, closed-circuit and surveillance television.

National Security Systems designs, installs, and maintains protective security systems for home, office, industry, and police departments—all are compatible wth cable tv.

Television Testing Company (jointly owned with Audits and Surveys, Incorporated) uses the capabilities of cable tv to test new television commercials or new product concepts and to offer market research surveys.

What other auxiliary services will help cable pay tv grow?

The microwave services will aid growth. Any of three FCC approved services will advance the development of both intracity and intercity cable television. Intracity communication services will include: the

AML and FDM/fm microwave systems, used to distribute pay tv programing between head ends within the city; and the Multipoint Distribution System (MDS), used to provide two service channels in the 2.160 to 2.172 GHz frequencies (in early 1973, frequencies were 2.160 to 2.170 GHz). The intercity common carrier services can interlink cities to provide operation from a single studio.

What microwave developments are on the horizon which will add to the viability of cable pay tv?

There are two major developments that are on the horizon. The first is the CARS Band Relay System, using digital modulation, operating in the 12.7- to 12.95-GHz band. A development between Laser Link Corporation and Northrop Corporation would make it possible to relay five channels in one area, interlinking all cable systems in the area. The many systems could then operate from a single studio and obtain the same quality of picture. By using digital modulation, it is possible to reconstruct the signal at each relay point. The signal would have the same signal-to-noise ratio at the end of the relay distribution, as it had at the point of origination. In present systems, the relay farthest from the source has more noise with respect to signal than those closer to the studio. Therefore, the same quality of picture cannot be relayed from area to area by the current state of the art. The Laser Link/Northrop proponents are advancing their development because they realize that the customer will not pay for a distorted picture. The performance of pay tv program relay drop to a system that may be two hundred miles away must be of the same quality as the service at the point of origination. The area shown in the drawing (Fig. 7-1) is that area where TheatreVisioN plans to expand its pay tv service from Sarasota. The expanded area will include all of the MSO's who have systems or franchises on the west coast of Florida. A survey shows that over twenty potential relay areas exist throughout the country, which justifies the need for this development.

Are there any developments on the horizon which will help the development of intracity service?

There is a development in the Multipoint Distribution Service (MDS) range (in previous answer) which could be helpful to hotel system proponents who are planning to use expensive studios with backup equipment. The MDS, which now has a 10-MHz bandwidth (which may be extended to 12 MHz), can normally accommodate two color channels, or one channel and 4 or 6 MHz of data service. One channel may be satisfactory for proposed services that plan distribution with

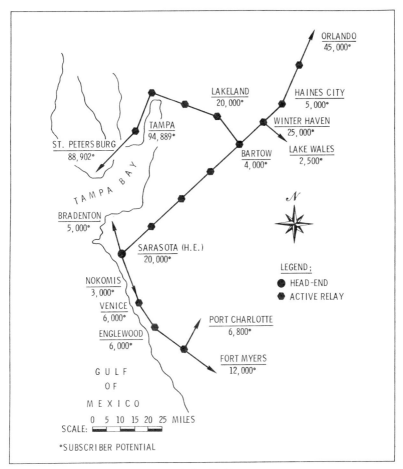

Fig. 7-1. Network for Florida west coast.

line-of-sight omnidirectional transmission and reception. However, there is a development, which uses the Sonic Vee patents, which will double the capacity of the MDS service. This Sonic Vee Corporation development, which belongs to Comfax Communications Industries, can provide four color television channels of service in the bandwidth assigned to two channels. It can also provide eight channels of black and white service in a bandwidth normally assigned for two color or monochrome channels. The technique, known as variable velocity scanning, has been successfully employed in the facsimile field. This narrow bandwidth technique is being built by Electronic Associates. Comfax is licensing others to advance this development for the MDS service. The development will enable cable operators to interlink their head ends with multichannel cable, while still permitting data transmission over unassigned bandwidths. MDS is a common carrier service and will be extremely valuable for pay tv signal distribution as the market develops. MDS services will be leased to transmit special events, on a committed schedule, to a number of systems within its receiving area. (The MDS transmitter is physically located by the common carrier group that has the frequency assignment from the FCC.) As pay tv develops, so will its commitments to MDS. There will be a greater need for bandwidth compression techniques, such as variable velocity scanning. Using these new advanced methods, more channels may be transmitted with higher quality.

What is the status of the MDS service as of the spring of 1973?

The basic microband service is operating today in Washington, D.C. under the leadership of a unique organization called the Microband Corporation of America, a subsidiary of the Arthur Lipper Corporation. They have also formed a national network, called Microband National Systems, for those who are planning to furnish this common carrier service under their corporate organization.

In the spring of 1973, Motorola announced plans to enter the hotel pay tv field. A single studio in each city will be used to distribute pay tv programing to hotels via the Multipoint Distribution Service. They plan to deliver programing in nine cities. Wells National Services Company, holding tv rights in 600 hospitals, has made a 50-market commitment with Microband National Systems. It is anticipated that a single studio may provide pay tv to all the hospitals within a given area at very low cost.

The MDS industry reports that there will be approximately 10 stations on the air before the end of 1973. There are more than 300 applications for MDS licensed stations in over 150 metropolitan markets on file with the Federal Communications Commission.

Which companies are furnishing pay tv equipment for the MDS field?

Encoding and decoding equipment is being offered by Digital Equipment Corporation of St. Petersburg, Florida, and the Laser Link Corporation of Woodbury, New York. Laser Link will have an operating system demonstrated in Washington, D.C. in May 1973.

Is it possible that a pay tv proponent could use MDS to deliver entertainment to residential subscribers?

Yes, such an operation would not be considered a CATV system under the FCC rules. These rules include as CATV systems only those facilities that receive and retransmit the signals of radio or television stations. Accordingly, the pay cable rules (which are part of the CATV rules) would not apply to such an operation. By the same token, the similar rules for over-the-air pay tv would not apply, since those rules apply only to broadcast stations.

The commission is well aware that, under its present rules, MDS is not subject to the restrictions imposed on the distribution of the same material by subscription cable or broadcast television. The FCC is presently considering the question of whether parallel regulation should be imposed on such distribution by MDS.

Is it likely that the FCC will impose regulations on MDS if it becomes a pay tv program distribution service to home subscribers?

It is the author's opinion that the commission will ultimately impose restrictions on MDS distribution to residential buildings (the same anti-syphoning restrictions as those for pay tv and pay cable). The distribution of feature films to hotels and other transient locations will not be subject to such restrictions. That is, in effect, the result the commission reached in the Sterling Manhattan-NY Telco case. It refused to interfere in film transmissions to hotels by telephone lines. If the phone company provided a similar service to residential buildings, or included sports programs for hotel viewers, the commission would "act to maintain [its] authority to protect the public interest." This is consistent with the commission's action in connection with the Trans-World Communications business service applications (which also proposed only the delivery of feature films and related material for distribution to hotel rooms). It granted the applications without imposing the 2- to 10-year anti-syphoning restrictions applicable to feature films, stating it would waive that restriction for any CATV system providing service to hotels in competition with a Trans-World fed operation.

What is the satellite potential of providing pay tv networking for cable systems?

In late 1972, the Canadian broadcast satellite, ANIK I, went into orbit. It is now broadcasting regular programs that blanket Canada and the northern half of the United States with reliable signals. The satellite distribution of tv will become a reality in the Rocky Mountain area in the spring of 1974. Many cable operators in that part of the country are now deeply involved with plans to establish ground stations. Received signals from the satellite will be microwaved or wired directly into local CATV systems. This planned program will provide two tv channels plus four audio channels.

Are there any plans for a nationwide cable television satellite?

Western Union expects to launch a communications satellite in the spring of 1974; it should be operational by summer of the same year.

Have any cable companies made any commitments at this date?

Yes, they have.

Have any pay tv companies made any commitments?

Not at the time of this publication; several are in the planning stages.

At this point, what will stop the pay tv proponents from leasing the video service?

The rate structure and the number of subscribers equipped with pay tv boxes will hinder service. Pay tv must develop on the ground first. There must be sufficient subscribers to justify the anticipated charges of the satellite service. The cable television market, at this moment, is small.

Has anyone estimated the cost of the cable operator of a domestic satellite service?

Yes. Hughes Aircraft, in its initial proposal to the FCC for a domestic satellite, estimated the cost to the cable operator would be $.25 to $2.00 a month per subscriber, depending upon the amount of programing.

What will make domestic satellite service feasible for CATV use?

The system will be more economical when urban CATV systems become larger and when ground microwave systems are more advanced. The degree of success of any urban cable television system will depend upon the availability of better films and special programing. When urban cable television develops, the satellite network of programs could become self-supporting. This is the only way that the small cable television markets can be stimulated to grow and eventually share in the satellite channel services.

Will the pay tv industry have to get together in order to take advantage of the satellite potential?

Probably, yes. Western Union, who will orbit the first satellite, envisions carrying data communications, but it will handle network television or cable television transmissions, if it can get such business. Perhaps, rather than waiting, all proponents could get together at some plateau of installations by 1975. There should be at least one-half million subscribers to pay tv programing by then. A deal might be arranged to pay a price of about one million dollars a year for a national pay tv channel. This cost could be shared, pro rata, by the pay tv proponents and the cable operators. Pay tv proponents could use the channel for certain specific hours, and the cable operators could use the channel during the nonpay tv periods.

What is the relationship between satellite reception and microwave?

Microwave will be used, in many cases, to distribute the signals between the satellite receiver and the CATV subscriber's head end. It is anticipated that new hubs of microwave distribution will be developed at the location of the satellite microwave receivers.

The Canadian government has funded use of television in satellites. Is the United States government doing any like funding?

Yes. The Rocky Mountain test, previously noted, is funded for educational programing and is being underwritten by government funds. The cable systems in the area are receiving these signals and distributing them, without charge, as a supplementary cable service.

Is direct satellite transmission to the home a practical service in the near future?

No. While satellite receiver prices might come down in future years, these receivers operate in a 2.5- to 12-GHz range and are too complicated and too costly for distribution to individual homes. However, even the smallest cable system could justify having a single receiver which converts programing to a useable channel. Assuming every cable system with 1000 subscribers had to spend $10 thousand, to provide the service, the operator would have to recover $10 for each subscriber receiving the service. If the service provided was a pay tv channel, the operator could recover his money in a short time. He would still have many hours to use the channel for educational or other programs he would like to provide to his subscribers. Obviously, for an individual home to acquire receiver capability would be totally unrealistic.

Is the future of cable pay tv dependent upon satellite networking?

Pay tv will not depend on satellite networking unless it becomes cheaper than terrestial interconnection. If cable groups participate in a CARS band area pay tv program distribution (described earlier), satellites may become useful. When satellite channels become economical on a per-subscriber basis, they will be attractive to the cable operator.

At the time of this writing, the pay tv proponents are meeting with the satellite advocates. When the time is right, pay tv service will be provided via satellite.

891756